All Rights Reserved

No part of this publication may be reproduced, stored or transmitted in any form or by any means, electronic, mechanical, photocopying, recording, scanning, or otherwise without written permission from the publisher. It is illegal to copy this book, post it to a website, or distribute it by any other means without written permission. Brief excerpts with full credit are permissible. If you use any of my material, please send me a brief note. I'd like to hear from you. My email address is on the final pages of the book.

Copyrights

I personally took many of the photos in this book. Other photos I purchased the rights to use. Some of the photos were listed as public domain on Wikipedia or Commons Wikimedia. When required, I added the owners name to his photo. However, there are a handful of extremely important photos which I included that did not seem to have any ownership or copyrights on them and were labeled for reuse without restrictions. So I included them. I would never knowingly use copyrighted photos or photos that had a watermark or logo on them. If I inadvertently used a photo that is copyrighted, I claim the FAIR USE doctrine for purposes of teaching, research and news reporting. The Fair Use doctrine clearly states that brief excerpts of copyright material may, under certain circumstances, be quoted verbatim or used for purposes such as criticism, news reporting, teaching, and research, without the need for permission from or payment to the copyright holder. With that said, if you should happen to be the owner of any of these crucial photos, please contact me and I will either give you full credit for the photo or remove it.

Note to Paperback Buyers

If you purchased the paperback version of this book, I would like to recommend that you have access to the internet while you read it. I added several links to my personal and private videos that have animations and science demonstrations which will give you a more complete media experience. I'd recommend being online and having those video links at your fingertips as you read the book. The links are on my website: stevelyke.wordpress.com

About the Author

Mr. Steve Lyke (pronounced LIKE)

I am not an Egyptologist, nor am I a scientist. I don't even have a college degree. Rather, I consider myself a visionary and truth seeker. And perhaps a prophet. At the turn of this century (year 2000), I could have cared less about the Great Pyramid of Egypt. My life was in total disarray and I was a desperate, needy man who was hopelessly trapped in a very dangerous situation. As a Christian, I looked up to the heavens for help and dedicated myself to work for God, if he wanted me. It was an unconditional surrender to Christ. I accepted my situation and my fate if that was God's will for my life. I offered myself to Him. Fortunately, God heard my plea and responded. He responded in a most unusual and unexpected way.

> And it shall come to pass afterward, that I will pour out my spirit upon all flesh; and your sons and your daughters shall prophesy, your old men shall dream dreams, your young men shall see visions: Joel 2:28 KJV

From that moment on, I was **overwhelmed** with dreams and visions. This was all new to me. Prior to that, I had no idea what a vision even was. Now they were coming in regularly. I knew that dreams and visions were in the bible, but my church didn't believe in those type of manifestations today, so I became an outcast with my church and my family. With God's Spirit heavily upon me, I heard inaudible voices that spoke encouraging words to me. And, the Spirit of God clearly wanted to teach me about the Great Pyramid. God wanted to use me and that knowledge as a part of His master plan. He wanted to start me off with the knowledge and secrets of the Great Pyramid. But it wasn't that easy.

I've spent many years of my life researching and studying the pyramid. I took two trips to Egypt. My DVD and book are a result of a lot of study and hard work. Thankfully, whenever I got stuck on decoding the pyramid, or just got confused, the Spirit of God would give me clues to help guide me or give me direct answers in dreams, visions or voices. It's an exciting spiritual story and I hope to publish it in the near future.

Truth Seeker Steve

I believe that it was during the summer of 2016 when I was in Hollywood walking down Hollywood Blvd just past Highland (near the Chinese Theatre) and I noticed a news van parked at the curb. Nearby, on the boulevard, a newsman was sitting down interviewing a woman. I looked at a nearby sign that read "Are You Superstitious?" Under the title it displayed the 20/20 logo from ABC.

They were interviewing people right off the street and asking them if they were superstitious and why. They wanted to hear what people were superstitious of. After watching a few moments of a woman's interview, I started to continue on with my walk. At that moment, a female representing the production company - perhaps one of the producers - asked if I wanted to be interviewed. I politely said, "no thanks... besides, I'm going to give a boring interview. I just don't believe in superstition". So I started to walk away.

That same production woman became a bit more interested in me. Her reply was something like this: "So you're a truth seeker then? I think that he (the male TV host) would still like to talk to you." I agreed to the interview. As I sat down next to the program's host, I told him exactly the same thing - which I'm going to be boring because I don't believe in being superstitious. Before the camera was rolling, he told me about some of the crazy superstitions that people have been telling him. I chuckled in disbelief. They may have recorded our conversation without me being aware of it. (They tend to do that).

Then the camera officially started to roll on me. In the past, I would have been extremely nervous at such an interview. I knew that this tape could be watched by millions of people. I was a bit uptight, but not as bad as I could have been.

As he asked me my feelings about being superstitious. The only thing that I can remember was saying that it **was all about science, and only science, no superstitions with me.** End of interview.

I watched and searched the internet for this tv program for months. I never found it. I assume that it never aired.

When i told my sister about the interview, the name of truth seeker resonated with her. She told me that truth seeker could me a new moniker for me. I liked it and thought that it was very appropriate. I decided to use it frequently.

Special Thanks

I wish to give a special thanks to Zuzana my production assistant in Cairo. Be sure to watch Zuzana's interview located in the video links section..or go to my website. Zuzana on the Giza Plateau

My first trip to Egypt in 2008, in front of the Great Pyramid of Khufu. This is where my dreams and visions became very interesting. After mounting the camel, I gave my camera to the Egyptian vendor to take this photo. He snapped it. I asked for one more  photo. He tried, but he said that the battery must have died. I looked at the camera and he was right. Batteries were dead. But how could this be?... I asked myself. The batteries were quite new. Not having any extra batteries, I had to leave the area without any more photos. A few minutes later, in my tour guides' car, I noticed that the camera batteries were sufficiently charged for more photos. My conclusion (which was verified by a vision in my hotel room), was that there still is an active electrical field around the pyramid that draws power which gets redirected up into the pyramid. From that moment on, I decided that the dreams & visions were valid and I needed to dedicate myself to this project.

The Sacred Mystery of the Great Pyramid DVD

It was during the year of 2009 when I produced my first Great Pyramid video and called it "Mystic Journey". (Listed on IMDB). During the subsequent years I added much more content and decided to totally reformat the DVD. I retitled it, "The Sacred Mystery of The Great Pyramid". While I have much experience as a 35mm photographer, 16mm film cinematographer and editor, I never shot a profession video nor produced a DVD. So needless to say that producing this video was quite a learning experience for me. I personally wrote the script, shot the photos, shot the video, edited the video, spoke the narration, created the animation and formatted the DVD.

Much of this book is a duplication of my video. Each format (book, Ebook and video) has something unique to offer. If you enjoy this book, you should also enjoy watching the video with the live action, animation and sound effects. In order to enhance this book and make it more of a complete experience, I added links to both videos and web pages. For the paperback book, my website will provide those links. Best to have internet available while reading the paperback. Website address: stevelyke.wordpress.com

There are many pyramids in and around Egypt. My studies, my book and my DVD are primarily focused on just one:

The
Great Pyramid

Contents

Chapter 1   Solving Earth's Greatest Ancient Mystery... Page 9

Chapter 2   Historical Retrospective .... Page 16

Chapter 3   Old School Belief .... Page 28

Chapter 4   The Pyramid Blocks Exposed........................ Page 38

Chapter 5   Wonder Stone of The Ancient Gods .... Page 49

Chapter 6   The Bottomless Pit .... Page 59

Chapter 7   The Energy Chambers .... Page 64

Chapter 8   The Fallen Angels .... Page 80

Chapter 9   The Fire Within .... Page 83

Chapter 10 The Golden Coffer .... Page 91

Chapter 11 The Electric Wind .... Page 97

Chapter 12 The War Between the Gods .... Page 105

Chapter 13 The Great Giza Circuit Board? .... Page 110

Links to Videos .... Page 117

Chapter 1
Solving Earth's Greatest Ancient Mystery

    The great pyramids of Egypt. Some have called them the most puzzling structures on planet earth. They have been referred to as an enigma, a riddle, a puzzle within a puzzle. Great men throughout history have come here to marvel upon them only to walk away confused and mystified. There are reasons why these pyramids defy explanation. If you're looking for relics of ancient Egyptian tombs, you won't find any.

    Most books tell us that the greatest of all pyramids, the **Great Pyramid** (above photo: hidden in back on right) is believed to have been built by the Egyptian Pharaoh, Khufu, around 2500 B.C.

Khufu was the second ruler of the 4th dynasty. He is generally accepted as having commissioned the Great Pyramid of Giza, but there is no evidence of it. It is also generally accepted that The Great Pyramid was a tomb, without any substantiated proof of it.

It is also a common belief that it took approximately 20 years to build and upwards to 50,000 slaves using copper tools to cut the stones in distant quarries and barge them up the Nile. Then, while harnessed to hemp ropes they dragged those massive blocks up long steep ramps and were lifted into place. (Keep in mind that this was before the invention of the wheel)! But, again, there is absolutely no clear evidence of this historical fantasy. Photo: The Great Pyramid of Giza

It is, however, very clear that the Great Pyramid is a structure that was precisely planned and constructed and that it probably couldn't even be duplicated today. Fortunately, more and more people are opening their minds to new ideas and rejecting the utter nonsense of what they have been previously taught. Let's take a closer look at the Giza complex...

Three of the very small Queens Pyramids upfront. Larger Menkaure Pyramid (215 foot height) behind them. Then the massive Khafre Pyramid at 448 feet. And in the rear, the Great Pyramid of Khufu (481 feet).

Walking up the Giza plateau toward the Pyramid of Khafre.
The Great Sphinx is guarding. Image Source: Hamish, Creative Commons

Pyramid Location: Cairo, Egypt. High atop the Giza Plateau

Great Pyramid is first pyramid on left.

Aerial view of massive Great Pyramid base. Cairo metro on right.
Google Earth

The Great Pyramid was built with an estimated

2.3 million stone blocks each weighing 2.5 to 15 tons each.

(Supposedly they were dragged & barged from quarries miles away)

Even if the workers had achieved the unimaginable feat of carving and piling up 10 perfectly shaped blocks on top of each other every single day, it would have taken them over 240,000 days or nearly 700 years to build! I'm afraid a bit too long of a wait for Khufu if he wanted a timely funeral service.

(Photo of author on massive blocks).

According to my math, I come up with a minimum of **275 blocks** that would be needed to be placed into position **every day** for it to be completed within 20 years! Totally impossible.

And if each block averaged 3 tons, 2 million of them would weigh a staggering 6 million tons! It's a man-made stone mountain!

450 ft. height, 756 ft. base square

Today, even after centuries of erosion, the Great Pyramid of Giza stands about 450 feet high. (The Statue of Liberty is 305 feet. The Washington Monument is 555 feet). The pyramid covers 13 acres of land.

Each side of the pyramid base is 756 feet long. (Over two football fields) When you consider its height of 450 feet, that results in a structure with a volume of over **85 million cubic feet!** That volume rivals the largest buildings in the world! The original World Trade Center Building contained 58 million cubic feet. (207 ft length x 207 ft. width x 1,350 ft. height). The Great Pyramid approaches the total cubic volume of the **two combined original World Trade Center buildings!**

Inside the Great Pyramid

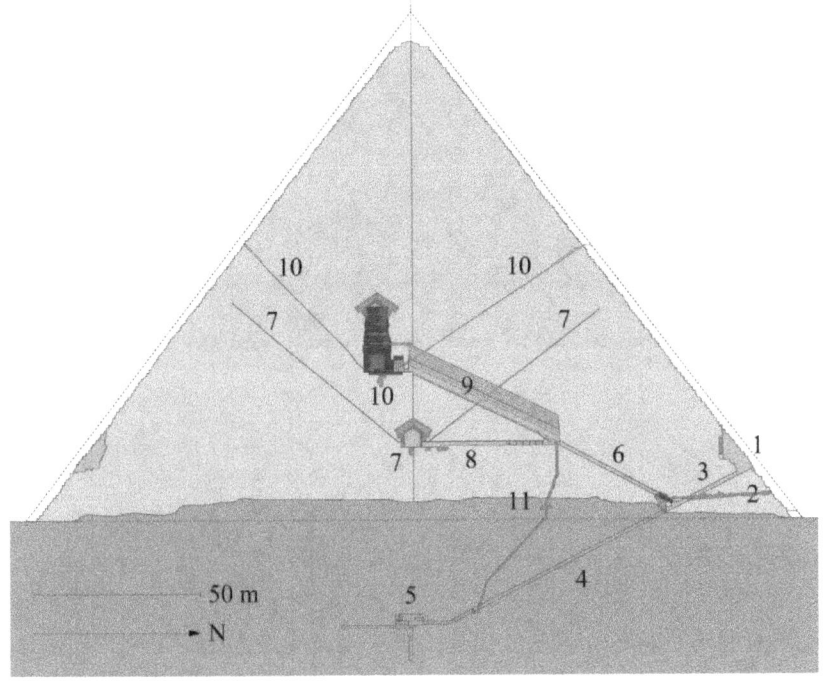

Wiki Media Commons Image

Diagram of Kings Chamber & Vents (10), Queens Chamber (7), Grand Gallery (9), and Subterranean Chamber (5). Current entrance and passageways to primary chambers (2) and (6).

So why such a massive construction effort with interconnecting chambers and galleries if you just need a tomb?

To truly understand why the pyramid was not recognized for what it really was, we need to take a trip **back in time.**

Rod Taylor in HG Wells, "Time Machine"

Chapter 2
Historical Retrospective

The Great Pyramid and Sphinx has been reported to have been discovered by the Arabs in 820 AD. Most of the Sphinx was **buried in sand**.

It is believed that during 820 AD, workmen for Caliph al ma'mun used battering rams to open up this entrance to the pyramid. It was said that no mummies nor tombs were found during his exploration. The only item found was a lidless "sarcophagus." (Today, this is the visitor's entrance to the pyramid.)

Boston Public Library photo

Centuries later, Napoleon's army invaded Egypt in 1798.

The Battle of the Pyramids, between French troops led by Napoleon and 21,000 Egyptian soldiers was a resounding victory for the French.

Napoleon develops a deep interest in the mysteries of the pyramids. He brings with him 154 scholars from every profession - from archaeology to architecture, medicine to geography, engineering and lexicography to survey, study and document them.

There were never any mummies found in the pyramid nor any evidence that it was ever a tomb.

He decides to spend a night in the King's Chamber. He refused to talk about it. And he forbid anyone from asking any questions about it.

It is later rumored that he had severe hallucinations and that he may have even seen his future and possibly his own death.

A few years later, a scientific breakthrough is presented to Napoleon. Alessandro Volta presents to him the first electric battery, known as the **Voltaic cell.** This revolutionary technology provided up to one hour of electric current.

B

Back in America, in 1816, Ben Franklin studies the **static electricity** coming from the sky. He refers to static electricity as "electric fluid" and "electric fire". While he wasn't the first to design a Leyden jar (to store electric current), he did group them together into a "battery" (a military term). He discovered that by multiplying the number of holding vessels, a stronger charge could be stored. Below left engraving: Franklin drawing electricity from the clouds in such a way as to be able to examine it, and prove that lightning is electricity.

Leyden jar battery acts as a capacitor, storing electrical charges from a static generator.
Franklin Institute photo

In 1850, Michael Faraday discovers that when an electrical charge is applied to water, **hydrogen gas** is released. It is called, **Electrolysis**.

Then around 1870, one of the most comprehensive surveys was made of the Great Pyramid by the Italian astronomer, Charles Piazzi Smyth. He entitles it, *"Our Inheritance in The Great Pyramid"*.

His incredible artwork and sketches gives us a closer look at many important details.

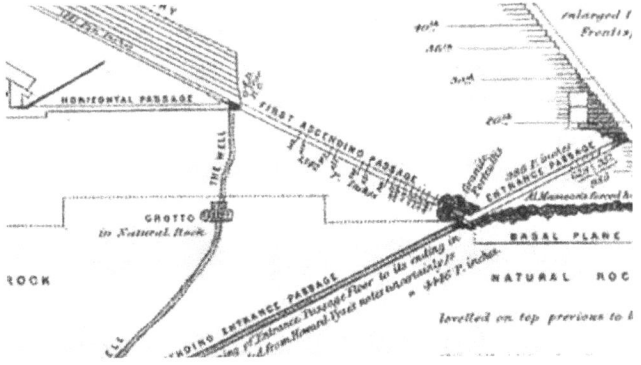

In 1879, the Wizard of Menlo Park, Thomas Edison, invents the first lightbulb. He puts on quite a light show at the Tennessee Centennial in 1897.

Electricity becomes available for cities.

Around 1880, the French Physicist, Pierre Curie, discovered that when you **compress** elements like crystals and certain ceramics, **Piezo electricity** results.

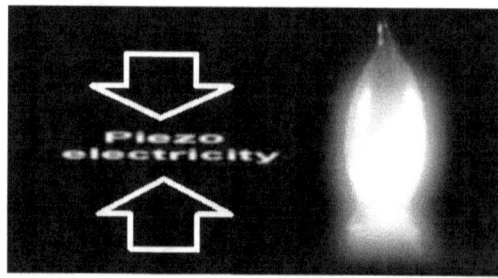

In the early 1900's, photographic cameras were getting more sophisticated and easier to buy. Priceless photos were taken in, around and above the pyramids. Much of the interior of the pyramid, along with the pyramid grounds, have been cleaned, renovated or has become unavailable for public viewing.

Note the "mastabas", or small tombs with the holes on top near the pyramid. Also note the creases in the pyramid surface on picture below. Will discuss later..

More priceless and revealing archival photos.

The Kings Chamber

The Coffer

The Grand Gallery

In 1917, Albert Einstein was the first to theorize about laser light. Right photo.

Forty years later, in 1960, Light Amplification by the Stimulation of Radiation (LASER light) was invented by Theodore Maiman. It utilized a synthetic ruby crystal.

Early 1960's Laser

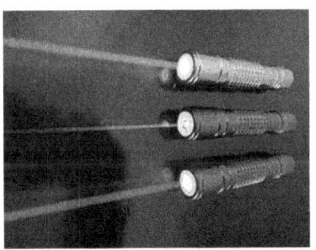

Todays' Laser Pointers

Then, in 1961, the first silicon integrated circuit chip was made by Robert Noyce in 1961.

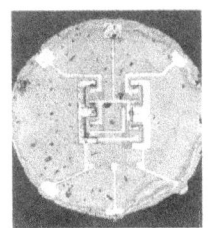

Photo courtesy: Fairchild Semiconductor International

The point that I want to make with my brief historical retrospective is this:

The technology used in the Great Pyramid, and the entire Giza complex could not have been recognized until our current age of high technology.

When the Great Pyramid was first discovered, there was no such thing as electricity. The study of the physical sciences was in its infancy. The sciences used in the pyramid were gradually discovered by man in increments over a 200 year period.

All of the science discoveries that I touched upon are an integral part of the real Great Pyramid. Without science and the work that it can perform, there would *not* be any Great Pyramid. (Photo reportedly taken in 1871)

A 16th century monk once proclaimed that:

"The Great Pyramid embodies all of the scientific wonders of the universe."

He was so right! But it's easy to understand why - over the millennia - it was considered only a tomb. Unfortunately today, it is still believed to be a tomb.

If we were to find a great pyramid on a distant planet, we would conduct every possible scientific test on it to determine its construction and purpose. And we would assume that extremely intelligent beings designed and built that pyramid. But back on earth, everyone assumes that the Great Pyramid was built by slaves, rather than by very intelligent beings. And that it was

built for only one person, a pharaoh and he wouldn't even use it until he died. Totally crazy!

I wonder how big it would need to be before the average person gazes upon the pyramid and says, "huh... a tomb? No way!" a) 1000 feet tall? b) Half a mile tall? c) ten miles tall?

- Just a hypothetical question -

Chapter 3
Old School Belief

As I touched upon earlier, common belief is that the Great Pyramid of Egypt is a tomb and only a tomb. And that this pyramid was built by tens of thousands of slaves during the reign of the Pharaoh Khufu (2575 BC). And that the slaves, using soft copper tools, cut and carved stones from quarries many miles away and were hauled up massive ramps and were completed within 20 years.

But there is no evidence to support any of those theories. There aren't any clear hieroglyphs documenting the quarrying, transportation, ramp building or lifting of the blocks. Not a single copper chisel has been found anywhere near the site. Nor is there any evidence of housing that many laborers.

We know that over 2 million blocks were used to construct the pyramid. Assuming that the masons worked 10 hours a day for one year, the math indicates that they needed to place 31 blocks in perfect position every hour. Or, **one block every 2 minutes to complete the pyramid in 20 years!**

If this incredible tale was true, the Egyptians would only have one method of lifting the blocks... ramps. Ramps which were composed of brick and earth that sloped upward from the ground to whatever height was desired.

But to carry an incline plane to the top of the 500 foot pyramid, would have required a ramp **5000 feet long and be three times as massive as the pyramid itself!**

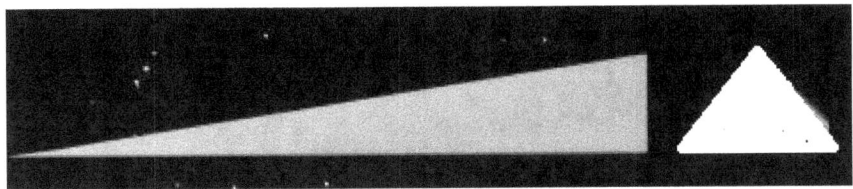

Furthermore, such ramps would have caved in under their own weight if they were made up of any material less stable than the blocks of the pyramid itself.

And what ever happened to the ramp and its remnants? An estimated **8 million cubic meters of surplus**! Where was it taken after the completion of the work? There is no evidence of it anywhere.

If the King's Chamber housed a tomb, where is the colorful artwork and ornate coffin? Both were traditions at that time.

This burial chamber is in The Valley of The Kings, 400 miles away.

And why would the King's Chamber need ventilation shafts? (10) And why are they on an angle? Horizontal shafts would have been much easier to build.

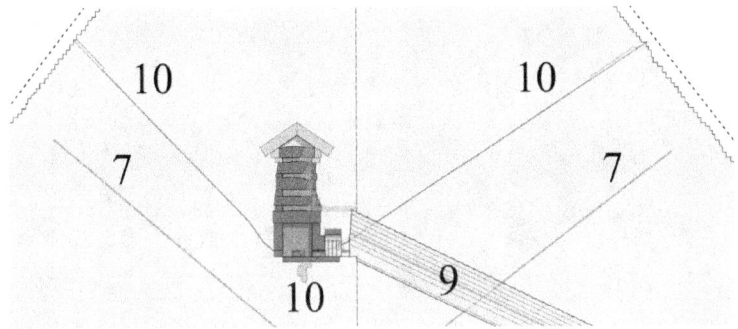

This shaft (10) that exits the King's Chamber, leads upward and outward - on an angle -toward the sky. It was once an entrance point from a source of extremely high heat as evident by the burn marks. Archival photo below also reveals that melting incurred. (Below photo taken before chamber was cleaned)

And why would there be a layer of thick basalt stone at the base of the pyramid? How could that possibly be for a tomb?

As I mentioned earlier, old-schoolers believe that the pyramid was built by Khufu, around 2500 BC. But the "Inventory Stella Text", discovered in the 19th century, indicates that both the Great Pyramid and Sphinx were in existence long before Khufu's reign.

Inventory Stella Text

Graham Hancock, author of the best seller, *Fingerprints of The Gods*, tours the Great Pyramid and makes the following observations:

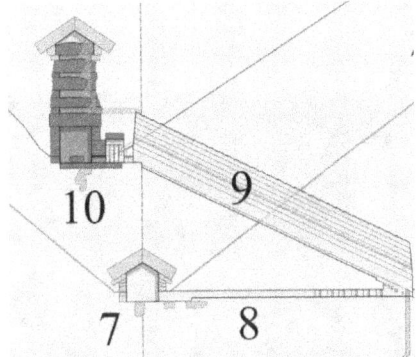

The Queen's Chamber: (7)

"Egyptologists have been unable to come to any conclusions about the niches in the wall, or even the purpose of the Queen's Chamber as a whole."

Note the square opening at bottom right. It's the channel (8) that leads to the Grand Gallery (9). Approximate dimensions of chamber: 18 ft. by 17 ft., 15 ft height.

The Grand Gallery: (9) "The most mysterious of all of the internal features of the pyramid. The gallery was required to support multi-millions of tons. Wasn't it not remarkable that a group of technological primitives designed such a feature 4,500 years before our time?"

Rails and steps were installed to help visitors walk the 26 degree incline in the 157 foot long Grand Gallery.

Above: Looking up the Grand Gallery (9) toward the entrance to King's Chamber (10)

Looking down toward the entrance channel to Queen's Chamber (where points 9 & 8 meet)

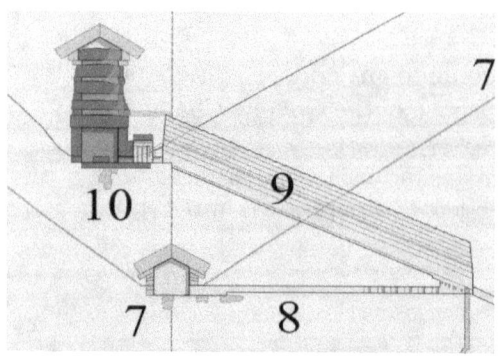

Graham Hancock writes: **"I was overtaken by the disorienting sense of being inside some enormous instrument of some sort. The primary impression that it made was one of strict functionalism. As though it was built to do a job"**.

The King's Chamber: (10): "It was a massive somber room, made entirely out of granite and radiated an **atmosphere of energy and power.**"

King's Chamber

Author: I would like to add that on my tour of the King's Chamber, I found it clammy, dank and a bit difficult to breathe. I didn't want to stay in there very long. I felt that I was in a machine that once ran very hot. It even smelled like a machine.

The Coffer inside The King's Chamber:

Graham writes: "Isn't it peculiar that at the supposed 'dawn of civilization', that the ancient Egyptians had acquired industrial-aged drills capable of cutting through hard stones like hot knives through butter?"   Author: Not very ornate for a coffin!

Rapid, machine-made spiral holes drilled into granite coffer

Graham concludes that **"nothing about The Great Pyramid makes any sense."**   Author: Makes absolutely no sense if you believe in those old school beliefs!

Joseph P Farrell, physicist and author of the book, *The Giza Death Star*, reveals that: "Beneath the Great Pyramid there are five massive stones or 'sockets'; four at each corner of the structure, and a fifth on the diagonal above the southeast corner.

These sockets are a 'ball and socket' joint familiar to modern engineering, permitting the building to rock and shift gently when the earth moves. This is the surest evidence, in and of itself, that the Great Pyramid is a coupled oscillator, for this feature is analogous to pressing down a key of a piano silently while striking another key to make it **resonate freely.** The pyramid, in short, **was designed to move.**"

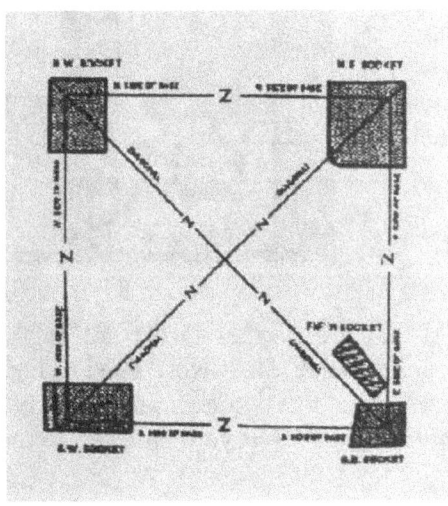

Chapter 4
The Pyramid Blocks Exposed

Natural stone vs. man made cast blocks

Mr. Moustafa Gadalla was born in Egypt. He is a certified Egyptologist. He has a degree in Civil Engineering and has authored ten books on Ancient Egypt.

In his book, Mr Gadalla writes, "We were taught that the pyramids were nothing but tombs built by tyrant pharaohs. And that slaves were used to haul up these big stones on temporary ramps. But when we examine the facts, **these beliefs are so incredibly baseless, that your faith in your education may be shattered.**"

"First of all, almost none of the pyramid blocks match the Giza bedrock chemically or mineralogically. The bedrock at the Giza Plateau is made of strata, while the pyramid blocks contain no visible strata (or marbling).

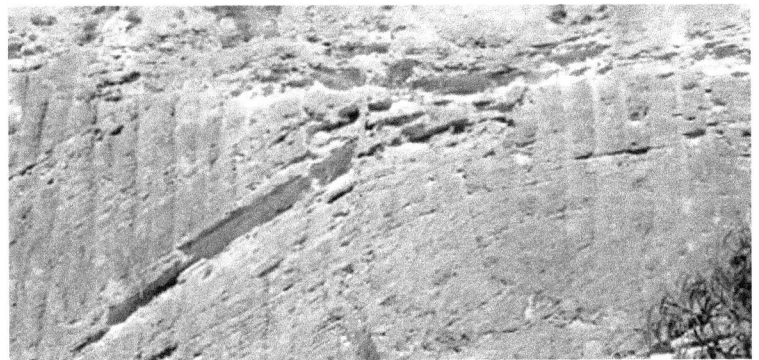

Above: Limestone deposit in Egypt contains strata

Below: close up of finished Egyptian limestone with marbling

Pyramid blocks contain no strata, no visible marbling

"Geologists and geochemists cannot agree on the origin of the pyramid blocks. This shatters old-school belief that it was quarried from local bedrock. And without modern cutting techniques, limestone frequently splits during cutting."

Limestone cracks easily

Limestone splits and shatters easily

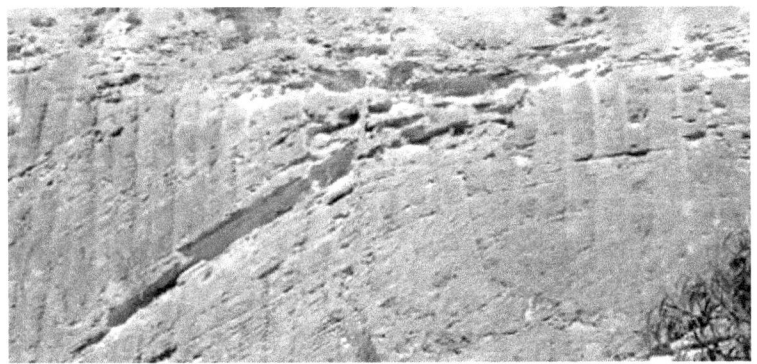
Above: Limestone deposit in Egypt contains strata

Below: close up of finished Egyptian limestone with marbling

Pyramid blocks contain no strata, no visible marbling

"Geologists and geochemists cannot agree on the origin of the pyramid blocks. This shatters old-school belief that it was quarried from local bedrock. And without modern cutting techniques, limestone frequently splits during cutting."

Limestone cracks easily

Limestone splits and shatters easily

"Faults in the strata bedrock assure that for every block cut - one will crack or be improperly sized during quarrying.

And yet note that the pyramid blocks were made so precisely, that you could not slide a piece of paper between them!

And given the many millions of blocks in these pyramids, there should be millions of cracked blocks lying nearby or somewhere in Egypt - but they are not!

Organic fibers, air bubbles and a red coating are visible on some blocks. All are indicative that the casting process was man-made and not natural stone.

The largest blocks found around Giza exhibit wavy, colorless layers and there are not distinct horizontal strata lines.

Furthermore, French scientists have found that the bulk density of the pyramid blocks are **20% lighter** than the local bedrock stone. Blocks which are manufactured are always 20% lighter than natural rock because they are full of air bubbles."
                              Moustafa Gadalla - The Pyramid Handbook

Close up of pyramid block with holes that probably originated as air bubbles.

Natural limestone is porous and very dry. One of the conclusions of the *Stanford Research Institute Electromagnetic Sounder Experiment on the Giza plateau in 1975 was that the Great Pyramid contained an estimated **100 million gallons of water!**

Clear confirmation that **water was added to make cast blocks.** (*Joint project between Stanford Research Institute Menlo Park, California, U.S.A. and Ain Shams University Cairo, A.R.E.)

As soon as I saw the below photo on the internet, I knew I had to include it in my book. It is **irrefutable proof** that the casting of pyramid blocks was taking place. The photo is a ground level block in front of the Great Pyramid which shows a distinct *crease and an irregular lip* at the bottom that would have been very difficult and totally **pointless to carve**. It's one of the "smoking guns" that I've been seeking. Photo printed with permission from Dr Michel Barsoum.

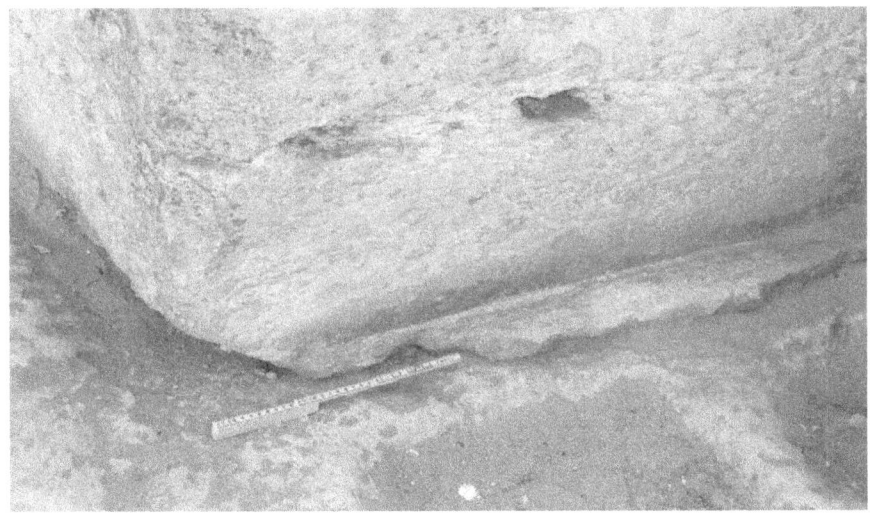

"I maintain that is it **impossible** for the Ancient Egyptians **to have carved this block, with such a fine delicate lip,** a few kilometers from where it is now sitting - right in front of Khufu - and dragged into place."

"The reason it is impossible is not because they were not capable of doing so, but that it would have been foolish to do, and **the builders of the Great Pyramid were anything but foolish.**" Dr. Michel Barsoum, Distinguished Professor, Dept. of Materials Science and Engineering, College of Engineering, Drexel University.

Dr Barsoum published a paper entitled, "The Great Pyramids of Giza; Evidence for Cast Blocks." Here are his thoughts and conclusions from this study:

"Four years ago, A. Ganguly, my graduate student, Dr. G. Hug, a colleague in France, and I, obtained some stone samples from the outer and inner casings of the Great Pyramid of Khufu. It took us 3 years, but we finally managed to ***prove beyond a shadow of doubt*** that indeed the inner and outer casing stones were **NOT natural.**"

"The inner and outer casings **are cast**. The backing blocks and the top halves of Khufu and Khafre are most probably cast.

The microstructure is consistent with a reconstituted limestone where the cementing phase is either ***silicon dioxide*** (silica is the major constituent of ***sand***) or a Ca-Mg silicate. One of the starting materials is believed to be *dolomite.* (an ingredient for the production of ***glass***).

In addition to being superb architects, civil and mechanical engineers, they were also ***brilliant chemists and material scientists.***

The fact that this artificial stone has not only survived for almost 5000 years but has ***fooled generations of Egyptologists and geologists*** is a testament to the incomparable genius of this ancient civilization!" (Italicized and emboldened words by author).

Link to Dr. Barsoum's Paper

Contrary to what we have all been taught, most of the pyramid blocks were not quarried. Instead, they are proven to be man-made cast molded blocks which **decimates all of the old beliefs about the pyramid's construction and the intellect of the pyramid architects.** This information alone should change the history books!

But this is only the beginning. I need to take this casting revelation much farther. I am not only convinced that the pyramid blocks were cast but also that they contained a unique blend of minerals that gave the pyramid a very special energy!

But first I need to talk briefly about sand....

Sand contains silica or silicon dioxide, the most abundant compound on earth. SiO_2

Today, silica gets highly refined and is used to channel electrical energy through silicon chips which are the brains inside all of our computers and electrical devices.

Silica sand is also used to make quartz crystal. **And quartz crystal has the vital scientific properties that the pyramid builders needed!**

But not all sand is created equally. The white sand found in tropical areas of the world contain large amounts of limestone.

Dark or black sands are rich in magnetite and basalt and have lesser amounts of silica.

However, the sands near the Giza Plateau are very rich in silica. Making the pyramid blocks with the silica rich sand will result in blocks unusually high in potential energy.

Just a few hours from Cairo, Egypt is Crystal Mountain.

　　Crystal Mountain contains some truly amazing quartz crystal formations. It's a geological footprint saying that the sands there are super rich in silica which would be ideal for use during the building of the Great Pyramid Laser Power Plant.

To the left and below: beautiful crystal formations in Crystal Mountain, Egypt

I'm convinced that these silica rich sands were used as an ingredient to manufacture **quartz crystal blocks.** While it may not have been clear and shiny, the blocks were quartz.

Quartz block

The hidden energy in quartz blocks would be used to accomplish the multitude of tasks that would make the Great Pyramid a total powerhouse! The evidence that I am presenting in the following chapters should make it crystal clear.

Chapter 5
The Wonder Stone of the Ancient Gods

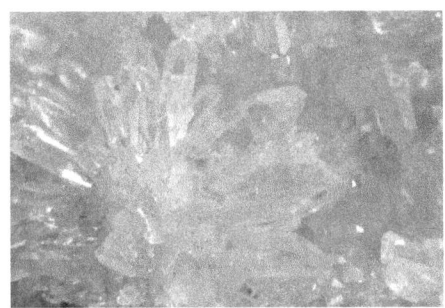

Natural quartz crystal cluster

S
Synthetic crystal
Photo courtesy National Ignition Facility

Quartz is a valuable and diverse mineral composed of silicon and oxygen atoms. SiO_2. Quartz is the second most abundant mineral in the Earth's crust.

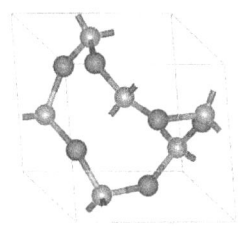

Crystal structure of quartz: red balls are oxygen, grey are silicon. Crystals are arranged in a highly ordered microscopic structure and are defined by their atomic arrangement within a space lattice.

Quartz crystals **amplify** energy, **transforms** energy, **stores** energy and **transfers** energy. They're used in diodes, oscillators, capacitors and are found in every circuit board. They're in computers, cell phones, lasers and satellites... just to name a few.

In the shape of a tuning fork, quartz crystals in this oscillator vibrate at a frequency of 32,000 HZ, or 32,000 times per second.

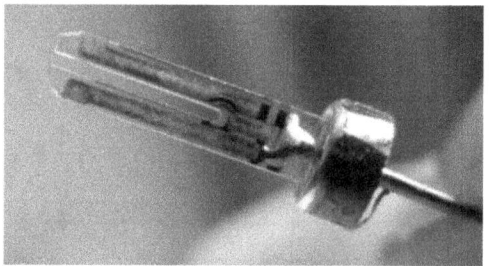

This precise oscillation makes the gear movement extremely accurate in wristwatches.

Quartz crystals have an exceptional musical resonant characteristic and can sing. This crystal bowl was made using a centrifugal mold. The bowl emits a powerful and pure resonant sound.

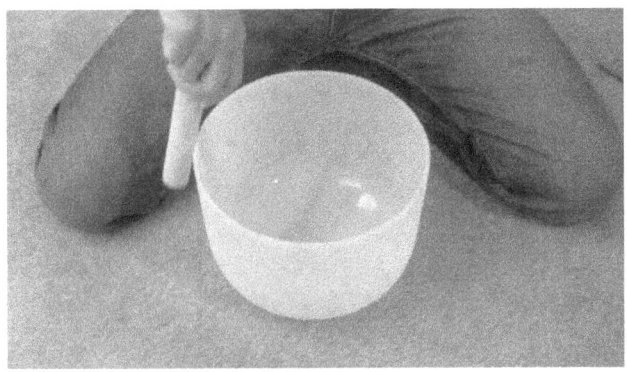

Singing bowls video clip

Quartz crystal bowls are made in all of the musical notes and the use of them goes back for centuries.

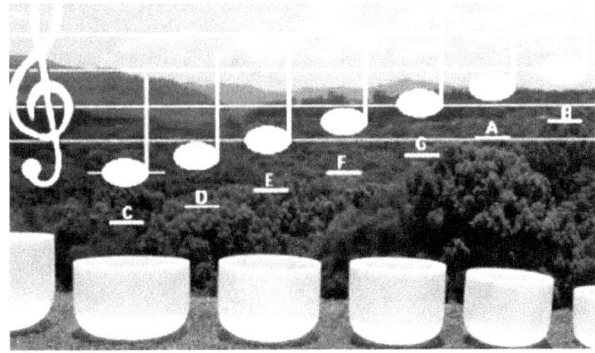

The sounds from these bowls were used for physical and spiritual healings. They're also used for relaxation and meditation. And others just love listening to the sounds that they generate.

Now *Piezo* in the Greek means *pressure*. As stated earlier, crystalline materials, such as quartz, creates piezo electricity when pressure is exerted on them.

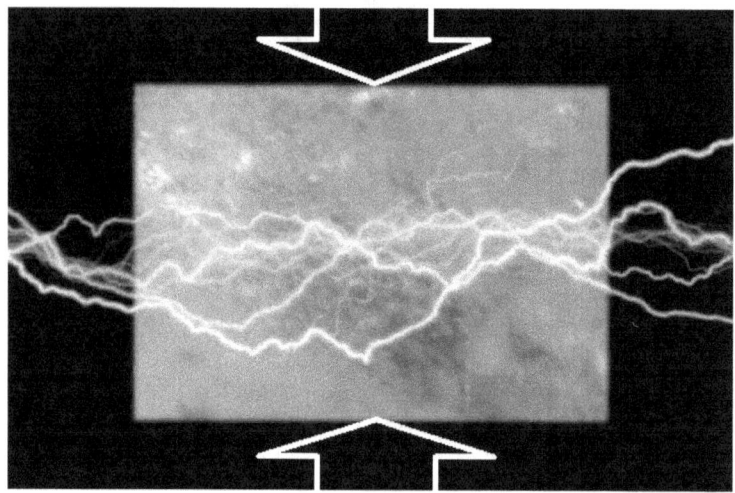

The popping noise that you hear when you pull the trigger of an electronic lighter is caused by a small spring loaded hammer hitting a piezoelectric crystal. This generates an electrical charge across the face of the crystal which creates a spark which ignites the fuel.

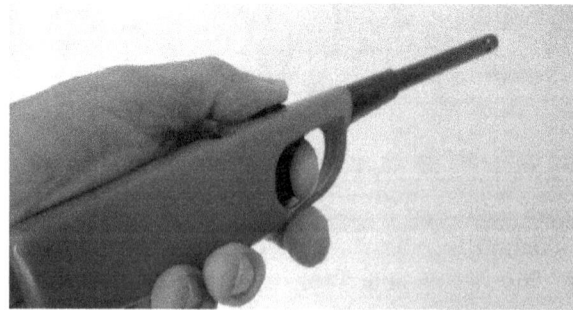

This piezoelectric demonstrator consists of a piezoelectric cell mounted in a retainer which allows pressure to be applied from a cam and lever. When the cell is compressed by pushing one side of the lever, **over 6,000 volts are generated** across the crystal and a powerful spark visibly jumps through the gap.

Wabash Instrument Corporation sold by Winsco

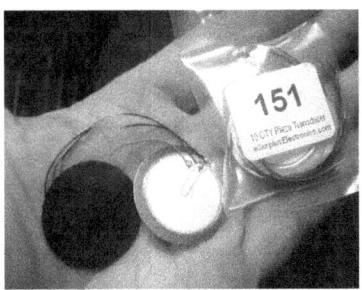

Sample size of piezo disc

When the lever is returned to level and the pressure across the cell is released, another spark, this time in reverse polarity will be generated as the cell returns itself to neutral. This is an example of the direct conversion of energy from one form to another: mechanical to electrical.

As previously mentioned, the silica rich sands near the Giza Plateau were used to manufacture quartz crystal blocks.

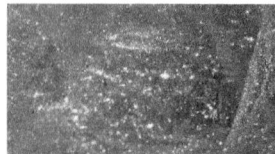 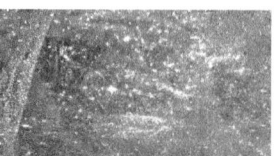 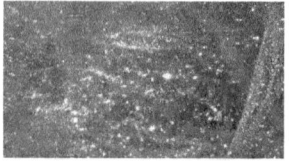

There are several ways to manufacture synthetic quartz crystal. But my focus is the way that it was made at the pyramid site.

I visualize a massive *smelting* or *foundry* type of operation with a very large furnace heated to extremely high temperatures. Sand, high in silica and other ingredients were being mixed, heated,

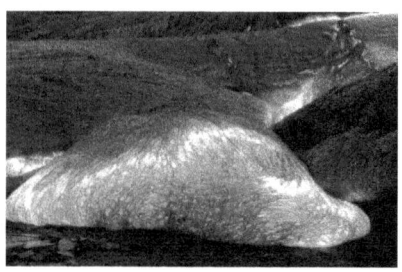

poured into a mold and placed into position while they were still soft *like molten lava.* Warm, soft blocks were placed upon cold and hardened blocks. **Joints between poured blocks would always be perfectly form-fitted** as the warm mixture cools and hardens against neighboring blocks.

Manufacturing blocks with cast molds results in absolutely **no wasted materials**. Any excess molten product which is not used or is spilled, can be saved and reheated for use in the next block. *This is why there are no block chips or remnants of any kind, anywhere near the pyramid.*

It's very interesting to note that quite a few blocks on and around the pyramid have large gouges in them. I have no idea how a mainstream Egyptologist can explain these indentations. I submit that a tool was used to maneuver the cast block while it was still warm and not yet fully hardened.

Dr Barsoum reminded me of a Sherlock Holmes quote on the art of deduction when solving a mystery:

"When you have eliminated the impossible, whatever remains, however improbable, must be the truth".

As a side note, I can't help to wonder if the King's Chamber was initially used as part of a furnace for heating and melting the sand. The chamber was built entirely from granite which wouldn't melt from the high heat. The chamber also had venting. Perhaps the granite coffer was used to pour the molten sand into?

It's possible that the King's Chamber (without the roof assembly) was the very first component of the Great Pyramid to have been built. Once the furnace was constructed, the builders proceeded with the casting of the blocks. Just a thought of mine.

Was the King's Chamber part of a furnace?

Was it the very first component to be built?

As I briefly touched upon in my historical retrospective, the importance of the electrolysis of water is to **produce hydrogen.** Electrolysis of water can be created by passing direct current from a battery (or electrical source) through water. By using conductive electrodes (like copper), hydrogen gas will bubble up at the negative side of the current and oxygen will bubble up at the positive side of the current. **If you add salt or brine, the reaction intensifies.**

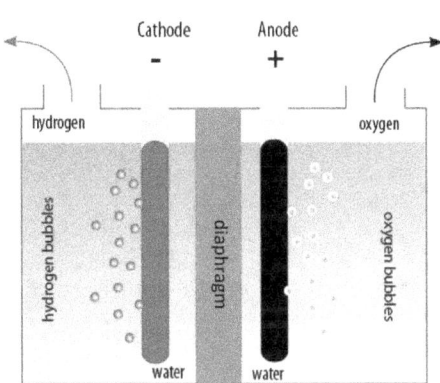

Please note that the hydrogen gas is extremely flammable! Don't conduct the below experiment at home.

Hydrogen balloons explosion provided by The Franklin Institute

Now let's step inside the pyramid and take a look at the so-called Queen's Chamber. Some old school thinking is that the pyramid builders used trial and error during construction and that the Queen's Chamber was a mistake! They say that originally it was going to be the King's Chamber but the builders decided that it was too small. Nothing could be further from the truth. It was designed and built for a *very specific function.*

The chamber, when first discovered, had a half inch coating of salt on its walls and ceiling. No other chambers in the pyramid had a significant coating of salt. (Charles Piazzi Smyth reported that it had such an extremely foul odor that it sent visitors quickly to the exits). You must ask yourself: "Why would one chamber have a significant coating of salt and have such a nauseous smell? The answer is that the salt in this chamber was a result of ongoing electrolysis. And the desired product of electrolysis is HYDROGEN gas. And hydrogen gas will be a key element which **ignites** this Great Pyramid Laser Power Plant.

"Hydrogen was important for the Great Pyramid to function"
Graham Hancock Blog

"Without hydrogen, this giant machine would not function"
Christopher Dunn, *The Giza Power Plant*

Chapter 6
The Bottomless Pit

Planet Earth is a giant heat machine. Our Earth provides a continuous flow of energy from its core to the surface.

Geothermal energy is that energy generated and stored in the Earth. In select areas of the world, below two miles from the Earth's crust, lies a pool of very hot water and steam. These areas are called geothermal hotspots.

These hotspots are usually below scenic treasures like Yellowstone National Park.

Hotspots run along the tectonic plates which are also responsible for Earth's volcanoes and earthquakes. These hotspots are known to be along the "Ring of Fire."

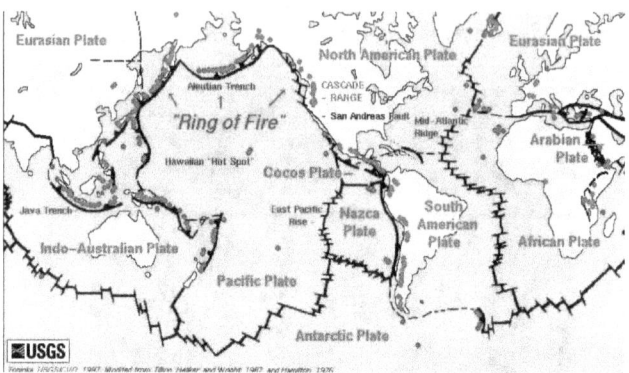

Tapping into these energy hotspots isn't a difficult task. For example, the US military has portable equipment which allows them to drill for these geothermal energy hotspots for use by its troops stationed near a battlefield. US NAVY Image

This geothermal power plant in the Philippines uses this natural energy resource to turn turbines which will provide 18% the country's electricity needs.

Geothermal power plants are traditionally built exclusively on the edge of tectonic plates where high temperature geothermal resources are easily available near the surface.

Now the development of **enhanced** geothermal technology allows energy extraction from a greater geographical range not limited by hotspots. Enhanced geothermal technology uses dry locations without any existing hot water or steam and uses high pressure water to open fractures into the 300 degree dry rock.

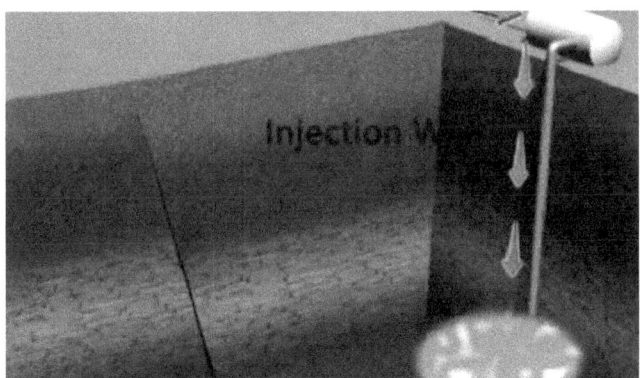

Then water injection wells are drilled to circulate water in this man-made reservoir. And then they extract the steam to the surface.

Now the country of Egypt, especially the area near the Great Pyramid, sits above the Arabian Tectonic Plate. It's a potential hotspot for geothermal energy, which I am sure did not go unnoticed by the architects of the pyramid.

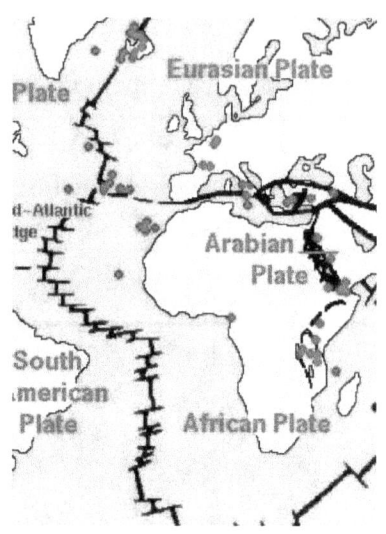

I don't think that it is any coincidence
that far under the Great Pyramid is a subterranean chamber. (lower left corner) which connects to the upper chambers.

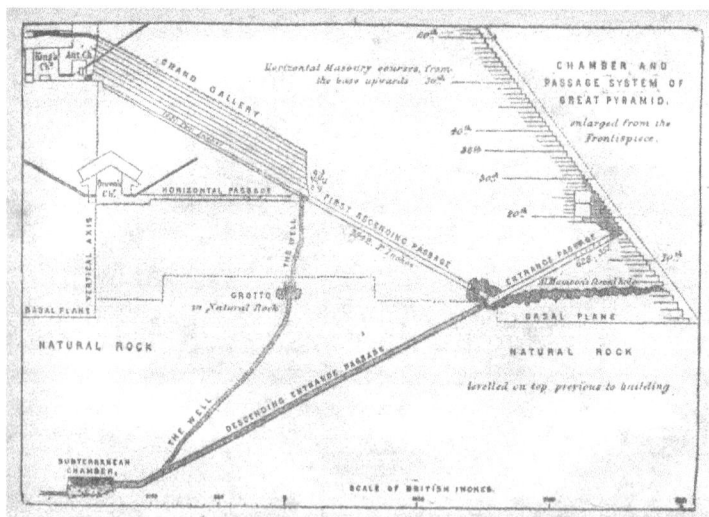

Inside this chamber is a well shaft that goes down to what has been referred to as "the bottomless pit". It was called the bottomless pit because no one knew how deep it was.

1910 photo

I am convinced that the engineers of the Great Pyramid designed this chamber to hold steam as they tapped into the geothermal energy below.

The steam would naturally make its way up to the Grand Gallery, where its heat would be needed for a hydrogen laser.

Whether they used an existing hotspot or created an enhanced hotspot, I am confident that the Earth's geothermal energy was used for the construction and operation of the Great Pyramid Laser Power Plant.

Chapter 7
The Energy Chambers

Let's take a closer look at the Grand Gallery. It is truly an intriguing chamber. It is 153 feet long (half a football field) and 28 feet high (from slanted floor to slanted ceiling).

Notice the corbelled walls. They get smaller by three inches on each side as you go toward the top. There are seven of these channels, so at the top it is only three feet wide. Cross section at right.

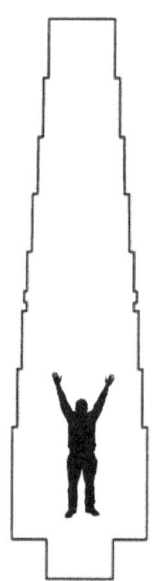

On each side of the floor shelf, there are 27 mysterious slots. A total of 54 slots.

Christopher Dunn, author of *The Giza Power Plant*, is one of the original proponents of a pyramid power plant. Mr Dunn has an interesting thought about the purpose of the Grand Gallery.

Dunn believes that 27 banks of Helmholtz resonators were in the slots of the gallery floor shelf.

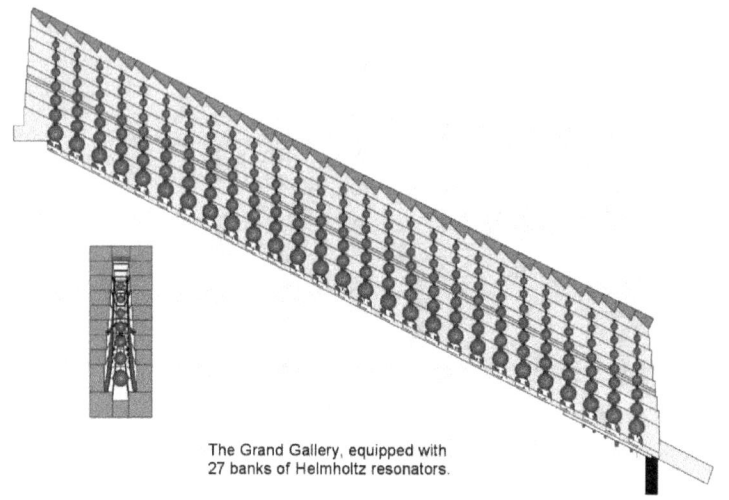

The Grand Gallery, equipped with
27 banks of Helmholtz resonators.

Dunn believes that these resonators were responding to vibrations coming from the Earth and was part of a some energy transfer. I have a hypothesis which I believe is much, much better. A little bit later, I will tell you exactly what I believe was in those slots. Right photo: Helmholtz resonator

But for now, let's walk up the Grand Gallery towards the King's Chamber. But before entering the King's Chamber, we must crawl thru a 2 foot by 2 foot opening to get into the **adjacent antechamber.**

Old school belief is that this chamber was built to keep grave robbers out of the King's Chamber. All modern observers firmly believe that any clever thief could easily climb over or around any obstacles in there. Like all the other chambers I believe that it had very sophisticated functions.

Two foot square tunnel from Grand Gallery into vertically notched Antechamber.

I believe that there was a notched granite slab which opened up (with pulleys) that allowed both light beams and sound waves to travel from the Grand Gallery into the King's Chamber. The notches helped to maintain a tight seal.

The Antechamber is a room that is 10 feet wide and 12 feet high. This empty room is equivalent to a modern day "plenum chamber".

Plenum Chambers

Plenum chambers are used today on high performance car engines to **equalize the air pressure** prior to going into the engine.

They can also be found in air ducts, acting as **acoustic silencers.** They help filter out the echo and unwanted noise that may travel from room to room.

The plenum chamber in the Great Pyramid accomplished both. It was built for **filtering** out unwanted echo & reverberation, and for **equalizing** the pressure inside the King's Chamber.

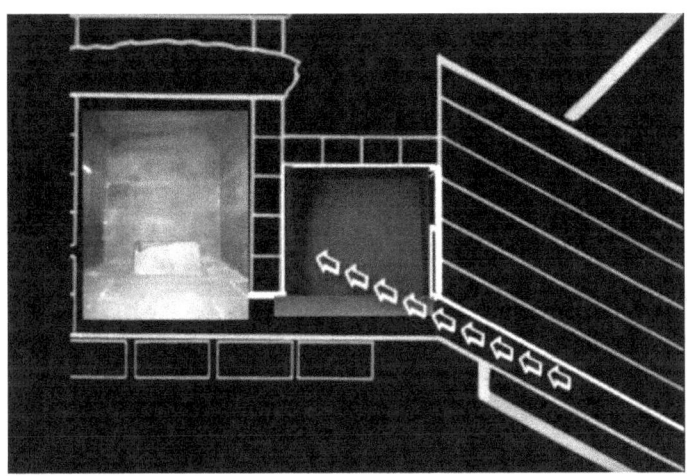

The importance of maintaining both pressure and vacuum inside the King's Chamber is evident by the super tight seals between the wall and floor joints. It was purposely constructed to make it impossible for air or gas to escape. (Hermetically sealed)

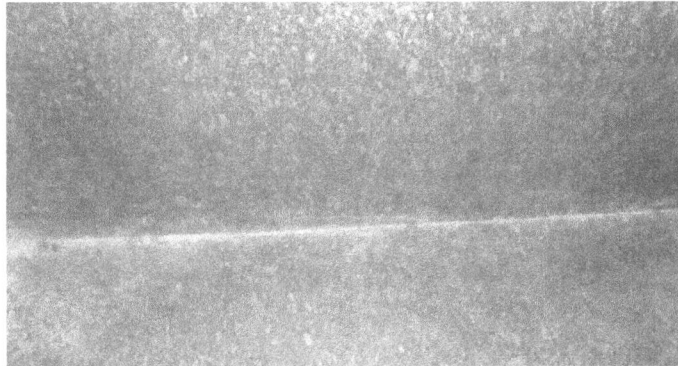

I believe that air pressurization and an alternating vacuum was regularly taking place inside the pyramid.

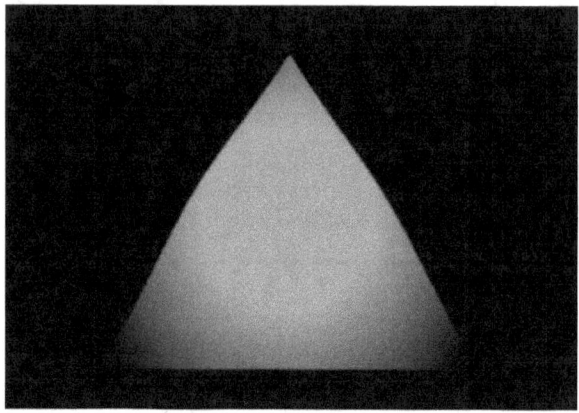

This compression and decompression stressed the pyramid structure. Evidence of this stress can be seen from this aerial photo. This compression and decompression fully activated the Piezo quartz blocks resulting in a continuous flow of electricity.

From the Antechamber you need to crawl again thru another two foot square opening in order to enter the King's Chamber.
> Right photo: Entrance from Antechamber into King's Chamber

OK, now you finally entered the King's Chamber. It is an absolute *masterpiece* in both design and construction.

Position of coffer has been moved away from its original position.

The **acoustic qualities** in the King's Chamber has been referenced and confirmed by numerous visitors since the time of Napoleon.

Napoleon's men discharged their pistols and noted that the explosion reverberated like "rolling thunder."

Another observation that early explorers made was that the sarcophagus or coffer was **tuned to a precise frequency.**

And, that even the King's Chamber itself was scientifically engineered to be a **resonator** of that frequency.

Paul Horn, a famous jazz musician and flutist, is considered the father of New Age music.

Mr. Horn brought a tuning device with him into the King's Chamber. It turned out to be tuned to the note "A". (438 cycles per second). He described the acoustic qualities of the chamber as *phenomenal with incredible resonance.* He went on to produce two music cds while playing his flute in the chamber.

Listen to Paul Horn play his flute inside the King's Chamber in my "Energy Chambers" video clip. Link provided in later chapter. Or go to my website.

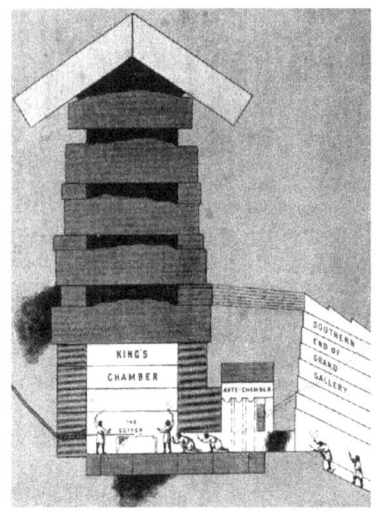 

Archival plate of King's Chamber

A cutaway model of the King's Chamber reveals hidden, yet critical components of its construction. Above the chamber there are five layers of granite beams. Each is separated by a space large enough to crawl into. The giant beams above the King's Chamber were clearly built to react to induced motion and **vibrate without restraint**.

Again, referring to Christopher Dunn's book, *The Giza Power Plant*, he observes that the giant beams above the King's Chamber are not built to relieve the chamber from excessive pressure from above but were **designed to fulfill a more advanced purpose.**

The layers were cut square on the bottom but were left untouched on the top surface which was rough and uneven.

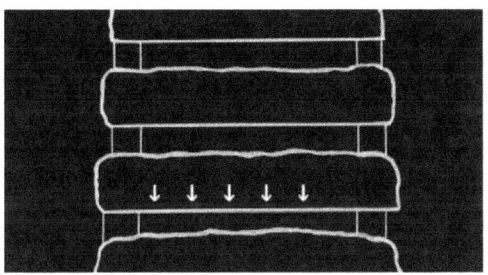

Some of the layers even had *holes gouged in them*. Early explorers called them bat holes.

It would be plausible to tune a length of granite by altering its physical dimensions (by gouging or chipping out some of its mass) exactly the way a bell is tuned - by removing metal from its mass.

Striking the granite beam, as one would strike a tuning fork would produce oscillation of the beam.

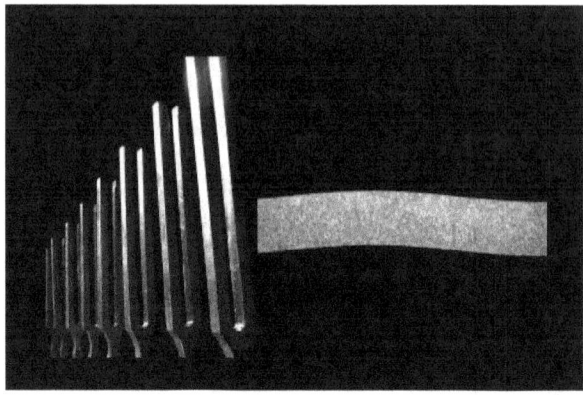

If we were to force just one of these beams to oscillate, with each of the other beams tuned to that frequency, the other beams would be forced to vibrate at that same frequency.

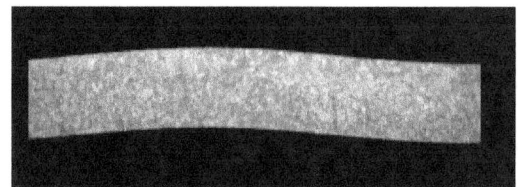

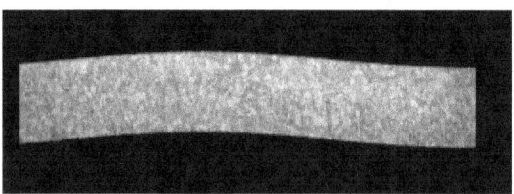

Tuning forks oscillation and resonance clip

So basically the beams above the chamber were like baffles in a speaker amplifying one specific frequency of sound. The angled roof above the beams returns all of the amplified sound waves right back through the beams and back into the chamber. Sound is amplified even more.

The granite flooring sits on a corrugated rock sub flooring. Sound waves that come down from the roof hit the sub floor which bounce upward and are *recirculated* which results in further sound amplification.

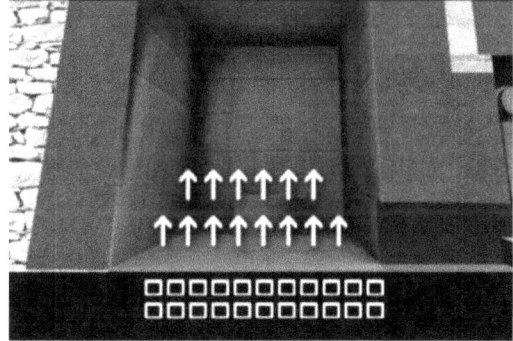

The King's Chamber now becomes an extremely powerful resonance chamber amplifying the sound energy which resonates throughout the entire quartz pyramid.
Resonance Chamber

As a parenthetical side note, inside the King's Chamber there's an "X" on the floor which marks the exact center of the pyramid. When you stand on the "X" and make a sound, or just hum, you can actually feel the sound reverberate and go thru the entire pyramid! And it seems like the sounds comes right back into your body! It was quite a moment for me to experience that. That alone was worth the trip to Cairo!

Photo by Andrew Currie

The Great Pyramid's massive sister pyramid (Khafre), along with other structures, would then resonate in harmony with the Great Pyramid. .

And since they were piezo electrified pyramids built high atop the Giza Plateau, they would generate pulses of electromagnetic energy which would be transmitted for hundreds of miles - perhaps thousands of miles - and be radio beacons into the universe.... And that's just the beginning!

Pyramid acted like a giant modern day radio tower

Be sure to check out the Energy Chambers Video clip!

Chapter 8
The Fallen Angels

So who exactly were the masterminds behind the design and construction of the Great Pyramid? I believe that the place to start the search is in the Book of Genesis, with the great grandfather of Noah, a faithful and holy man named Enoch.

"And Enoch walked with God:

and he was not;

for God took him."

Genesis 5:21-24

We are told that the prophet Enoch walked faithfully with God. And that God took him into the heavens and that he never experienced death as we know it. (Enoch's mysterious translation is repeated in Hebrews 11:15)

Enoch's ancient scroll, later referred to as the **Book of Enoch**, reveals the story of the fallen angels which was revealed to him while in heaven.

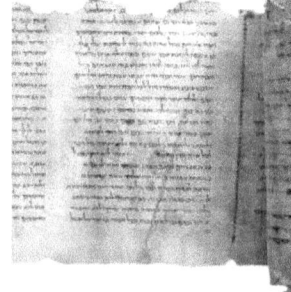

But Enoch's incredible story was considered **too controversial to be included** in what we now call the Holy Bible. (I believe it was rejected at the Council of Nicaea 325 AD.)

In his forbidden scroll, he wrote that 200 angels rebelled against God. And that these fallen angels were cast out of heaven and down to Earth. Painting: "The Fall of the Rebel Angels" by Luca Giordano.

These fallen angels found the female inhabitants of Earth desirous and had sexual relationships with them. Their offspring was known as the **Nephilim.**

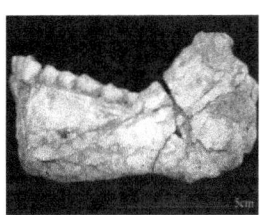

As far as I can see, there is no indication exactly when the fallen angels were cast out of heaven. I personally believe that is was *very early in Earth time.* (Photo of the oldest fossil remains of Homo sapiens. It's dated as far back as 300,000 years ago.) The first almost complete adult mandible discovered at the site of Jebel Irhoud. Photo by Jean-Jacques Hublin, MPI-EVA, Leipzig.

The Enochian Scroll tells us that the Nephilim were half angel and half human - and some were **giants.** (These giants are mentioned numerous times in Enoch's scroll. I believe that they were as tall as the treetops, perhaps even taller.) These giants can also be found in Numbers 13:33, "All the people whom we observed were giants. We also saw the Nephilim, the descendants of Anak. Compared to the Nephilim, **we're like grasshoppers."**

But more importantly, Enoch also says that they had **God-like intelligence**. He writes that one of the fallen angels taught man how to: **"make swords, shields and the fabrication of mirrors and the use of stones."**

So the fallen angels and their Nephilim offspring (with God-like intelligence) were masters of metallurgy, chemistry, physics, mathematics and astronomy (just to name a few) when they met up with the first human inhabitants on Earth.

There was one particular species of the Nephilim that masterminded the Great Pyramid complex.

And I, the author, **saw both the image and name in a vision. They were called the**

"Voltaic".

They are the ones that built the Great Pyramid. And I believe that this happened many tens of thousands, perhaps even **hundreds of thousands of years ago.** Long, long before Adam and Eve.

Chapter 9
The Fire Within

Pyramid: In Greek, Pyr means "fire", amid means "within".

Forty miles east of San Francisco, in Livermore, CA. is the home of The National Ignition Facility. Larger than three football fields, the NIF is building a complex that will house 192 laser beams and when complete, it will be the world's most powerful laser.

One fourth of their floor space will contain over 3000 glass amplifier slabs. These amplifier slabs will be used as a gain medium which will amplify the laser light which will duplicate fusion reactions similar to those that power the sun and the stars.

NIF fusion lasers

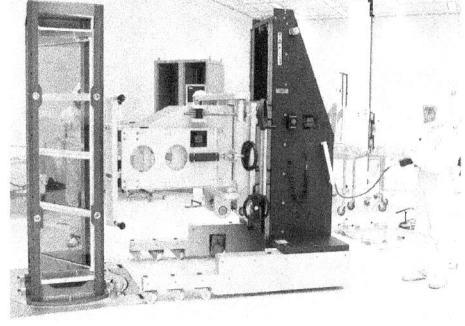

The ancient Voltaic gods also built a facility that utilized the **power of light**. Let's go inside the pyramid and take a look at the world's first laser.

The Queen's Chamber is the master oscillator room where the laser light is born. First, the quartz blocks create the piezo electricity which energizes the chamber.

Partially filled with salt water, the current goes through the brine and produces electrolysis, resulting in hydrogen gas which creates the laser light. (It's possible that other catalytic elements were added into this chamber through its ventilation shaft).

The newborn light then travels through a charged channel as it makes its way to the Grand Gallery.

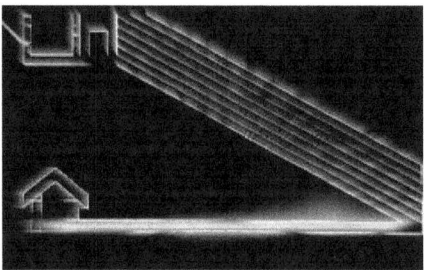

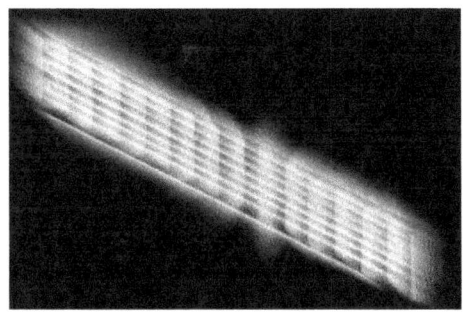 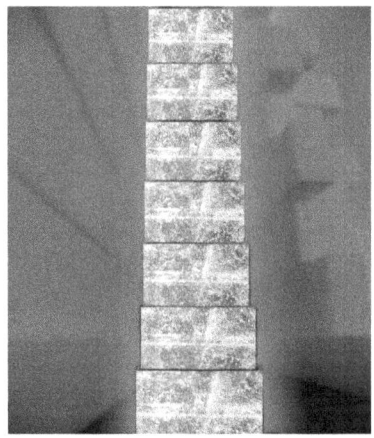

Tuned, synthetic crystal amplifier slabs were housed in the mysterious channel slots in the Grand Gallery..

They were positioned to receive the laser light on a 56 degree angle to reduce reflection and light loss.. This angle is known as "Brewster's Angle." (exact angle not shown in animation) Brewster's Angle (also known as the polarization angle) is an angle of incidence at which light with a particular polarization is perfectly transmitted through a transparent dielectric surface, with no reflection. Gas lasers typically use a window tilted at Brewster's Angle to allow the beam to leave the laser tube.

The Grand Gallery now becomes a laser cavity with light bouncing off mirrors through columns of crystal amplifier slabs where a chain reaction takes place where photons continually collide with gas atoms to create a powerful beam of laser light.

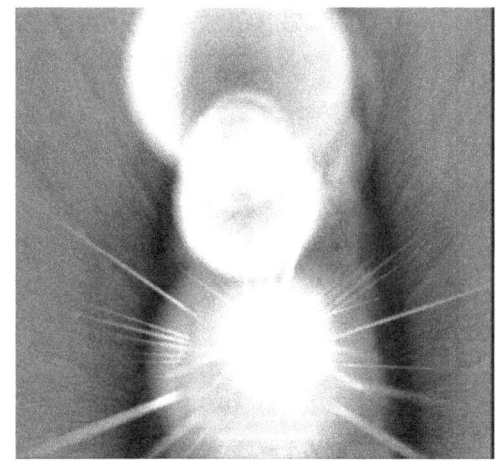

The sliding granite slab in the antechamber acted as a laser shutter. It was closed to keep the young light from escaping and then opened when the light was sufficiently amplified.

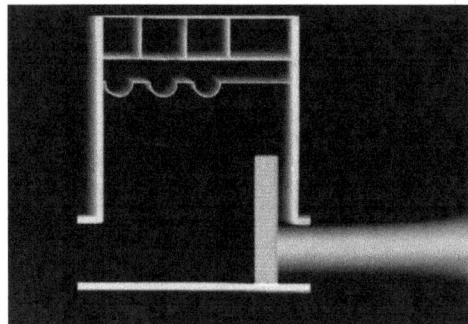

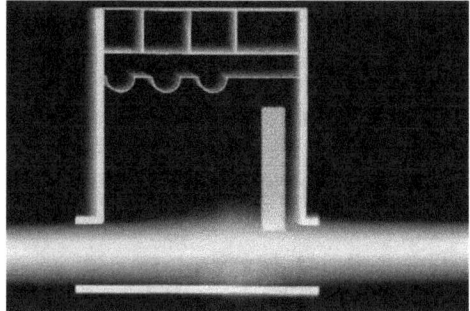

The intense laser light enters the King's Chamber, where a beam splitter divides the light into two separate beams. The primary beam enters the northern shaft.

And out emerges a laser light so powerful it enters the heavens where it seeks the pulsating waves of energy from the stars.

The sun and the stars resonate... you can actually listen to their pulsating waves of energy.

The entire universe sings out like a multitude of musical instruments. I call it the **symphony of the stars.**
Symphony of Stars

Scientists can record the sounds of stars. The technique, stellar seismology, is becoming popular because the sounds give an indication of what is going on in the stars' interior.

The pulsating sound waves from the stars then ride our **laser's return trip** from the sky to the pyramid. (*See Video Link on Laser Sound*). Those sound waves enter back through the ventilation shaft and then return into the Grand Gallery.

Now there are seven courses or channels that run the entire length of the 150 foot Grand Gallery. The top width is three and a half feet. Each lower course adds six inches until it reaches a maximum width at the bottom of seven feet. Our laser cavity now doubles as a sound resonation chamber...

...with each channel tuned to resonate the frequency of the seven musical notes.

After travelling through the antechamber (plenum chamber) the King's Chamber receives the sound from the Grand Gallery and **amplifies** it.

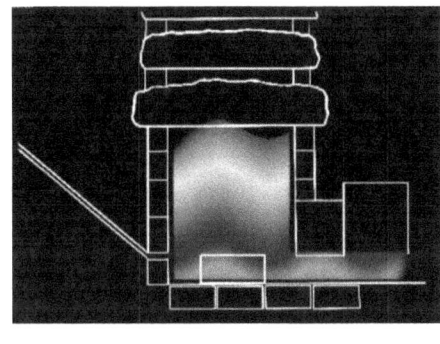
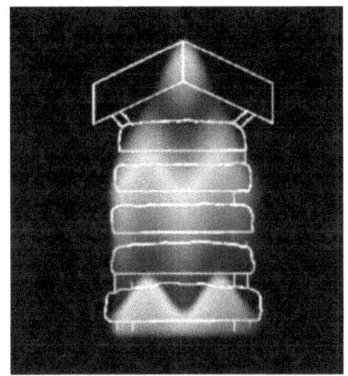

The roof assembly above the chamber becomes a sound **attenuator** to smoothen out sound wave fluctuations.

Now on the floor of the King's Chamber, we have the granite coffer. On the top of the coffer are three machine-drilled holes.

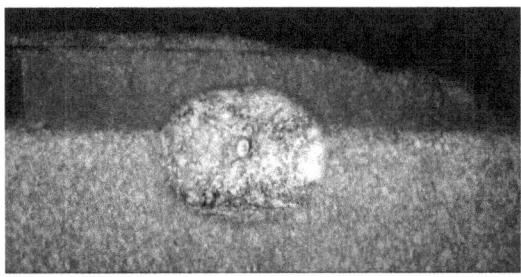

Circular inset showing close-up of drilled hole

These drilled holes tightly held small nuclear fusion pellets. The intense laser energy of the laser beam activated the pellets.

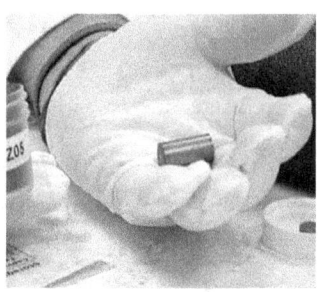

Modern day pellet above

NIF fusion target below

Which resulted in a nuclear fusion burn... just like that which occurs naturally in the sun and stars.

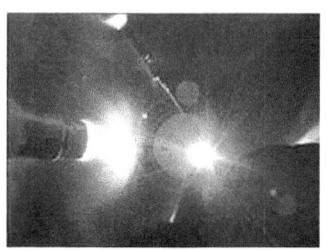

The multi-purpose King's Chamber now also becomes a containment structure for a **nuclear power plant** emitting extreme heat capable of producing steam and providing power for additional electricity production.

Note: Nuclear reactors don't need to be large. Nuclear power fuels this mini sub.

Chapter 10
The Golden Coffer

The Philosopher's Stone - a legendary substance which transforms a base, like mercury into gold.

I often wondered about the purpose of the coffer inside the King's Chamber. In particular, why it looked like a rectangular bathtub made of granite. It puzzled me for years.

I was watching the TV program, *Ancient Aliens*. The episode was entitled, "Aliens and Temples of Gold". Then it finally hit me.

At one point the narrator asks, "Was gold made in the Great Pyramid?" Michael Dennin, PH.D., Professor of Chemistry, University of California, Irvine responds:

"When you think about how gold is made... naturally within the nuclear reactions of the sun. Then it spreads throughout the universe when a supernova occurs. And the sun explodes and sends all the products out.

Because we can now do controlled nuclear reactions, we have the ability finally to essentially manufacture gold from other elements."

In 1924, 150,000 volts of electricity directed into mercury isotopes produced gold.

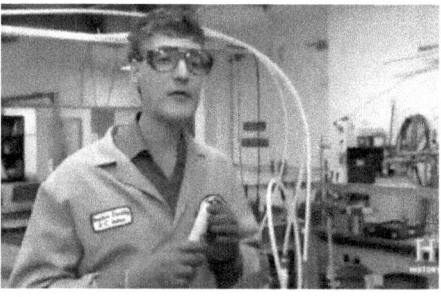

The narrator continues, "Here at the University of California, Irvine, Dr A.J. Schacka conducts experiments in alchemy every day." Dr Schacka explains, "We put mercury in one of these tubes. **Mercury has an isotope that will actually pick up a neutron and in about 23 hours turn into gold.**" (After extremely high voltages of current run through it.)

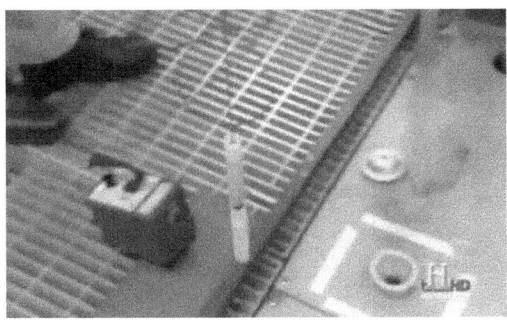

Don't try this at home because the cost of the electricity is greater than the value of the gold that it creates.

The narration continues, "A real life Philosopher's Stone at work in the basement of a university building."

Ancient Aliens Making Gold Clip

Mercury plus electricity makes gold!

With staggering amounts of electricity flowing into the King's Chamber, the **coffer was used to make gold**!

Gold is the most valuable element in the universe. Gold has many unique properties. Gold is **malleable.** A single gram can be beaten into a sheet of one square meter (10 sq. ft.). Gold is **ductile**. One gram of gold can be drawn into a thin wire 165 metres long (500 ft) and be just 20 micrometers thick. (.0007 inches). Gold is **non-corrosive**. It never reacts to oxygen which means that it will never rust or tarnish. And, of course, it's been used for coinage and jewelry. Ancient civilizations also used gold compounds for medicinal purposes.

Most importantly, gold is a critical element in both electronics and in space exploration.

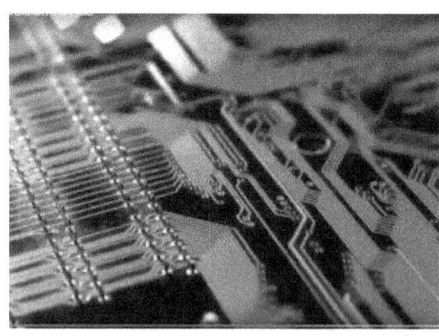

By far the biggest industrial use for newly mined gold is in the fabrication of corrosion-free electrical connectors in computers, cell phones and other electrical devices.

Experts agree that there was an abundance of gold in Egypt. But where did it all come from? Silver is typically found in conjunction with gold. Yet silver in Egypt was not abundant. In fact, silver in Ancient Egypt was *highly prized and rare*, especially when contrasted with the abundance of gold. It is believed that Egypt had to trade for its silver. How can this be? To make it more interesting, there were copious amounts of mercury. Vials of mercury were found in many ancient tombs.

Tut's golden mask

Above: A gold ring featuring ducks bearing the name of Ramesses IV. Louvre-Egypte

Left: golden sarcophagus

The gold was manufactured in the coffer inside the King's Chamber by converting the element of mercury Hg to the element of gold Au.

Mercury is much more plentiful, but less valuable than gold. And mercury deposits are at or near the surface of the Earth making it easier to mine than gold. Photo: Sulfur mercury mine in CA.

Mercury is made by crushing cinnabar and heating it. Cinnabar has been mined for thousands of years. Even as far back as the Neolithic or Stone Age which began about 10,000 BC.

Cinnabar

Early cinnabar oven

The crushed cinnabar was placed in ovens which were used to collect mercury vapor. The mercury vapor, when cooled, became pure mercury.

The Voltaic were **masters of alchemy.** They used the extracted mercury to manufacture gold inside the Great Pyramid. Ancient Egypt, as we know it, (3000 BC), inherited the manufactured gold from the Voltaic.

Chapter 11
The Electric Wind

The wind... for millennia, it has been recognized as a powerful energy force. An energy force which was used to help mankind accomplish the task at hand.

Today, this free energy force is still being harvested.

But there's a similar energy force which isn't as obvious. It's an unnoticeable, undetectable type of wind that I believe was harvested by the Voltaic to accomplish their tasks at hand. Created by static electricity, it is known as **electrostatic induction,** I call it the Electric Wind.

The knowledge of static electricity goes back to the earliest civilizations. But for centuries it was just a mysterious and interesting phenomenon.

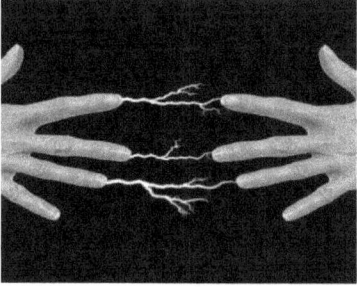

Above left: Benjamin Franklin *"Drawing Electricity from the Sky"*, painting by Benjamin West

As I mentioned in a previous chapter, during the 18th century, static electricity generators were built to create electrical energy. The electric current would be *stored in a Leyden jar*, which was an early form of capacitor.

1883 Wilmshurst generator with Leyden jars

Many science museums have exhibits displaying the effects of static electricity.

In the early 1900's a higher voltage static electricity generator was designed by Robert Van degraaf. During 1929, he developed his first generator producing 80,000 volts, by 1933 he constructed a larger generator generating 7 million volts.

Today, we can buy a simple science kit that generates small amounts of static electricity which creates the electric wind and demonstrates its effects.

My handheld Van de graaf stick uses an insulated belt which spins past contact brushes on both sides. It separates and accumulates ion charges. You can see the byproduct of the electric wind (spinner acceleration) which is being created solely by static induction. There is no wind coming from the nozzle.

These aluminum cupcake pans turn into flying saucers that defy gravity and levitate. Many UFO enthusiasts believe that this is the propulsion system behind many UFO's.

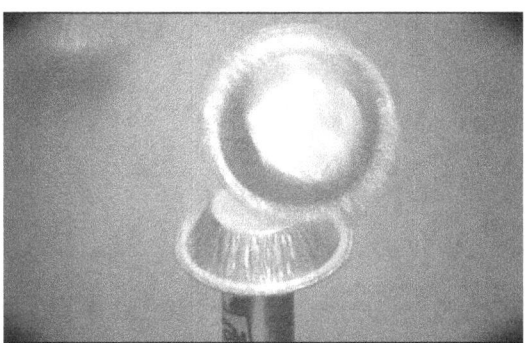

My small handheld static stick will actually charge the gas inside this neon lamp.

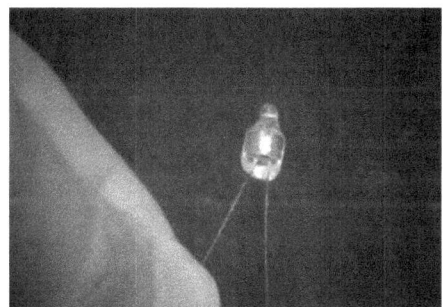

The gas inside the lamp creates plasma. The process that makes the lamp glow is the same process that is in a plasma TV.

The Great Pyramid was a massive electrical generator. A natural byproduct of this electrical generation was static electricity... lots of it. And the science wizards behind the Great Pyramid power complex put it to work. With their knowledge of physics and electrostatic induction, there was very little that they couldn't accomplish.

While they probably didn't have plasma TV'S, they could of have designed glass lamps or orbs which would have provided them with light at night or light for their temples.

Glowing electric lights in temples would have been an impressive sight, demonstrating the power of the gods and priests to the humanoids on Earth.

"If you want to find the secret of the universe, think energy, frequency and vibration." - Nikola Tesla

Nikola Tesla is considered the all-time most popular electronic wiz. He is known as the father of high frequency, high voltage electricity. His revolutionary Tesla Coil created extremely powerful electrical fields. They have been known to light up fluorescent lamps up to one mile away without any wires.

In the photo above, Tesla holds a gas filled, phosphorus coated bulb which was illuminated without wires by the electromagnetic field generated by his coil. Tesla envisioned a time when his towers would have provided unlimited wireless electricity to the public.

*Tesla claims to have received 3d **visions** of blueprints which helped him.*

He also envisioned a day when his high frequency wireless electricity would be transmitted from tower to tower through the ionosphere, which would result in the lighting up of the oceans. (The ionosphere is a layer of the Earth's atmosphere that contains a high concentration of ions and free electrons and is capable of reflecting radio waves).

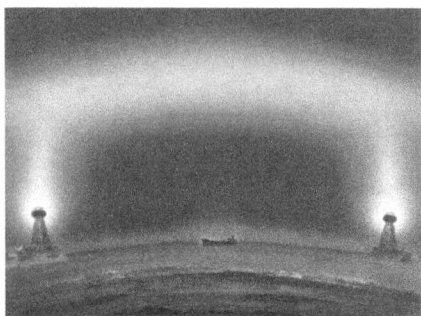 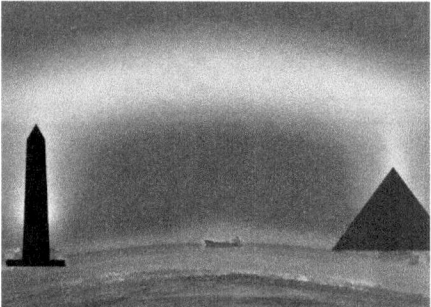

Tesla's vision on left, author's vision on right, with obelisk receiving transmissions.

While the Tesla experiments are intriguing and demonstrate a unique wireless form of energy, I believe that the Voltaic successfully harnessed wireless energy. Not as Tesla envisioned, but rather with the *Electric Wind*.

During peacetime, both water and air navigation utilized the electronic rivers in the air created by the pyramid which also extended into the ionosphere to other structures - similar to Tesla's vision.

Chapter 12
The War between the Gods

So what happened to the pyramid's quartz crystal blocks and how did they turn to blocks of limestone?

As I stated earlier, I believe that the Great Pyramid is tens of thousands, perhaps hundreds of thousands of years old and may even go all the way back to the beginning of known time. It's very possible that the quartz blocks just slowly decayed over that extremely long time period. I don't think anyone really knows how manufactured quartz crystal would look after 500,000 years of exposure to the harsh plateau elements and a worldwide flood.

Regardless, I am personally convinced that there is another reason why the blocks look and test like limestone today.

I believe that this region was devastated by a hydrogen bomb. Perhaps a great war took place among the various factions of fallen angels.

A Hydrogen bomb can reach one million degrees in a split second. Merely all matter at the center of the blast is vaporized. Extremely high levels of radiation are at the center of the blast.

These blocks, on the opposite side of the main entrance to the pyramid cleary show that they melted, burned and almost vaporized from a punishing heat. It takes 2,950 degrees Fahrenheit or 1,200 degrees Celsius to melt stone. Looks like these blocks were deforming as the immense heat turned them into a molten lava-like substance. This side of the pyramid took a direct hit from the blast. Blocks closer to the ground don't look as bad as these mid-level blocks.

Color enhanced photo shows what they probably looked like during meltdown.

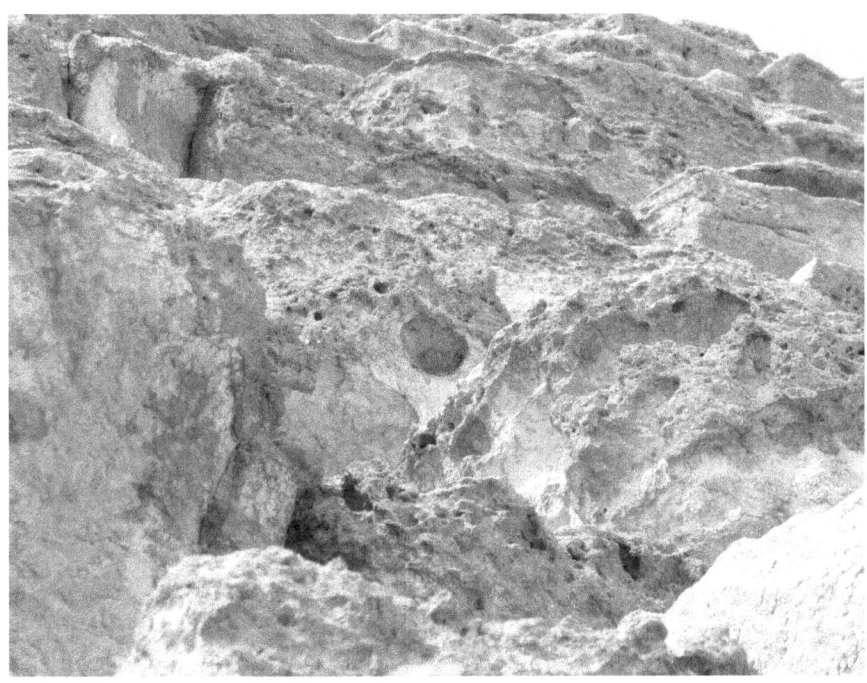

Close-up of melted and fused mass of pyramid blocks. Can't even see their shape anymore.

But when the devastation was over, the hot quartz crystal blocks got totally re-molded. As they cooled down, many cracked and were transformed into their current state of frozen, shapeless, petrified blocks of sand with a limestone look.

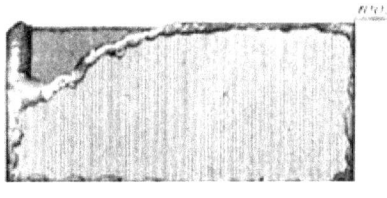

Archival plate of granite coffer shows evidence of an internal meltdown also.

During the hydrogen blast, the pyramid was bombarded by electromagnetic radiation: microwaves, terahertz waves, infrared waves, along with *ionizing ultraviolet rays, X-rays and gamma rays! *ionizing of atoms causes chemical reactions which totally destroyed the crystalline structure in the quartz blocks. It was **LIGHTS OUT** for the pyramid!

The once mighty quartz blocks of the Great Pyramid no longer had the magical scientific characteristics that they once enjoyed. (However, I believe that deep into the layers of ionized blocks, there remains some that were protected and may still have a little bit of a charge left in them.)

While it's probable that the Great Pyramid laser power plant was made primarily for peaceful purposes, its laser would certainly have destructive power. And I believe that its power was ultimately used for war.

Navy laser weapons system US Navy Image

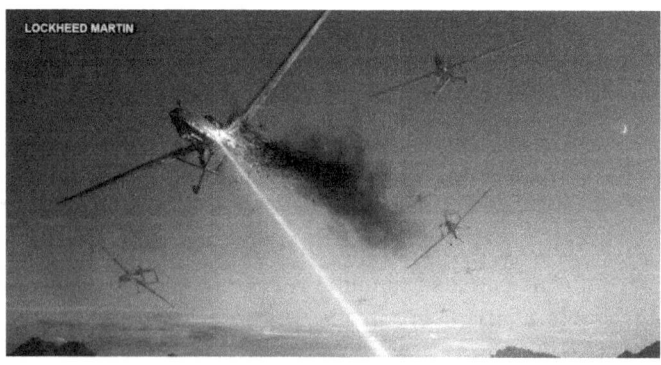

Lockheed Martin Image

It's quite interesting that the legend of Atlantis speaks of a highly advanced civilization that possessed **super crystals**.

Legend tells us that they ultimately used **their technology for war** and that Atlantis was buried in the sea.

After studying the amazing technology in and around the Great Pyramid, along with all of the destruction, I can't help to think that it may have played a significant part in this legend.

Perhaps Atlantis wasn't buried in the sea. Instead it was buried in a sea of sand?

Note: I know that Plato, the philosopher, wrote that Atlantis was an island in the middle of the Atlantic Ocean. But Aristotle believed that Plato, his teacher, had **invented** the island to teach philosophy.

Chapter 13
The Great Giza Circuit Board?

When I initially glanced at the very first integrated chip that was designed, I couldn't help to be amazed by the similarity it had with certain aspects of the Giza complex. Then I looked at some aerial photos and was dumbfounded at the similarities the complex had with a circuit board.

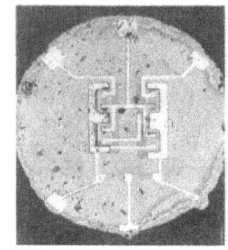

And I am not the only one. In an episode from the TV program, Ancient Aliens, entitled, "Aliens and Ancient Engineers." The narrator states, "From the air, the layout strangely resembles a circuit board, with two large processing chips."

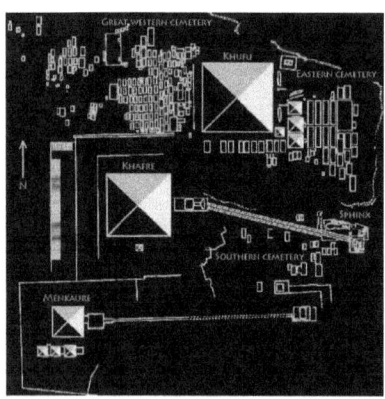

The main components to a circuit board are: the battery (or

other supply of current), resistors, transistors, capacitors and diodes. **All control or direct the flow of current.** These components are assembled on an insulated green fiberglass base.

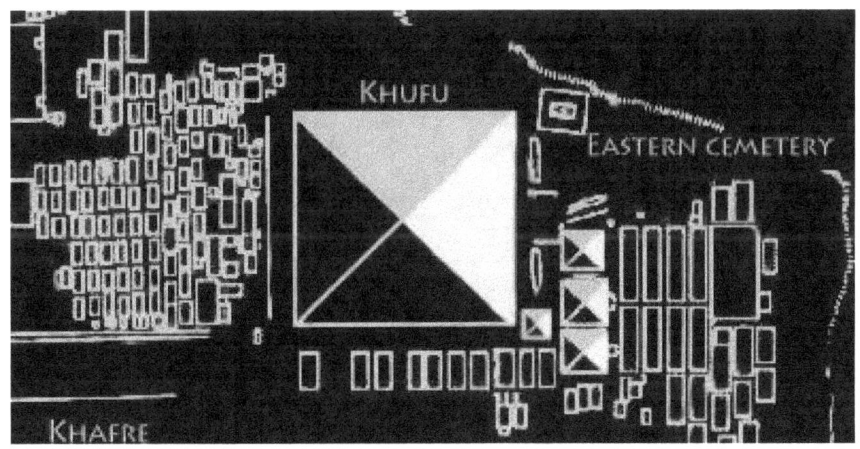

Could it be that the so called, "mastabas" (left of pyramid) actually once acted like transistors or diodes on a circuit board?

Is it possible that the three small pyramid structures in the "eastern cemetery" (known as The Pyramids of Queens) acted as Leyden jars or capacitors to store electric current?

And perhaps that's why there's a foundation of basalt volcanic rocks around the Great Pyramid - to act as a insulator. Just the way the green fiberglass acts as an insulator on circuit boards?

Layer of basalt rock acts as an electrical insulator preventing "short circuits" and to absorb the heat on the Giza circuit board.

Is it possible that all of the structures on the Giza Plateau - including the Sphinx - worked together as a giant memory system, or computer?

Did this computer guide spacecraft?

Did the computer or Artificial Intelligence have robots programmed to be protectors of the galaxy?

Is this what the future holds for us?

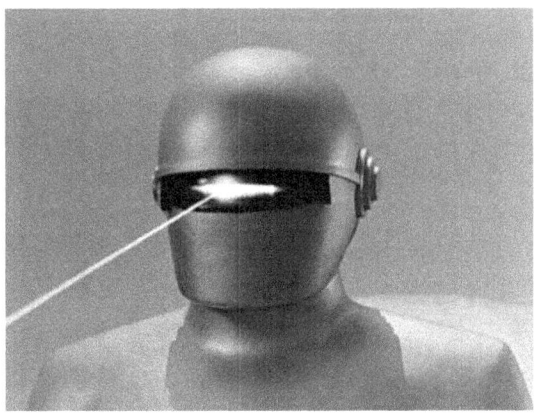

1951 The Day the Earth Stood Still 20th Century Fox

Or worst - A future invasion of robots programmed to steal and possibly kill??

Do I honestly think that this invasion can happen? YES!

A few months ago, I had a powerful early morning dream. I looked in the darkened evening sky and saw the dimming sun with another large orb coming from behind it. The orb was coming fast. As it approached, it became faster and faster. It was another planet coming directly toward us. I found myself saying a loud and fearful, "OH NO.!", which woke me up. (Hope my roommate didn't hear my Oh No shout). This dream clearly was about the mythical Nibiru planet that the ancients wrote about in their texts. I had a vision of this planet several months ago, so this is my second spiritual communication I received about it in recent months.

This gets me to a book that the Spirit of God wanted me to read some 14 years ago: *The Twelfth Planet* by Zecharia Sitchin. The book is described as follows: "The 12th Planet brings to life the Sumerian civilization, presenting millennia-old evidence of the existence of Nibiru, the home planet of the Anunnaki and of the landings of the Anunnaki on Earth every 3,600 years, and reveals a complete history of the solar system as told by these early visitors from another planet."

It seems that the ancient Sumerians believed that Planet X was inhabited by a superior race of space travelling beings called the Anunnaki. It is also believed that the Anunnaki came to Earth in search of gold, which it needed to restore the depleted atmosphere of their own damaged planet.

The "Nibiru cataclysm" is a feared disastrous encounter between the Earth and a large rogue planetary object which will take place in the early 21st century. Believers in this Planet X doomsday event refer to this object as Nibiru. Sitchin & Planet X

While I don't believe that Planet X will hit Earth, I do believe that a close encounter could be catastrophic.

Since I had my "OH NO" dream a few months ago, I decided to also tell you about a radio broadcast that I heard of a man who claimed to have worked for the CIA. He said that he was one of only a few individuals who were able to look thru the Hubble telescope while it was focused on this Planet X. Either this guy was the world's best actor, or he did actually see this incoming menace, along with its long trail of destructive trash, coming our way.

His narrative was extremely detailed, passionate and very convincing. During the radio interview he convincingly claimed that Planet X was definitely heading our way and that for sure, it would create major havoc with our planet earth as it approaches and bypasses us.

One more dream for your consideration: In the dream, I was watching uniformed robots, who looked and spoke exactly like our males and females. But they were definitely robots. They had a job to do. Their purpose was to steal our wealth. They were confiscating our gold, jewelry and other valuables. They were going door to door. They did it without any emotion. Earthlings seemed totally helpless or just lacked the will to fight them.

Did the Anunnaki have a war with the Voltaic? Were the Anunnaki responsible for the hydrogen bomb blast which destroyed the Great Pyramid's Laser Power Plant?

Will the Anunnaki come back to visit us when their Nibiru Planet has the next close encounter with Earth? And will they bring robots to steal our wealth??

We have been taught that mankind and his technology evolved very slowly and has gradually progressed through the ages along a straight line path.

But the technologies of the ancient gods uncovered in this book prove that tens of thousands of years ago, perhaps hundreds of thousands of years ago, an extremely advanced civilization once inhabited this Earth.

Video Links
Must be connected to internet
Links are also available on my website

Energy Chambers Video

Laser Sound Demo

Interview with Zuzana

The excellent link below was independently produced:
Piezoelectric Crystals: Effects & Demo

UFO at Great Pyramid

stevelyke.wordpress.com

Author's email: Truthseekersteve @ Gmail.com
Please understand that I cannot answer any questions.

Just A Few More Things...

For the longest time I couldn't decide about the UFO incident at Roswell... you know, the story of the crashed ship and aliens that were found. I just finished watching Dr. Steven Greer's video documentary entitled, "Unacknowledged". (He also has an excellent book with the same title). Several retired military men came forward and acknowledged that it happened. All gave the same basic account of it.... 3 to 4 feet aliens with big eyes died in the crash. One of them survived for a while before he died.

So now I come forward and state that I believe that it all happened and that there is an ongoing government suppression of UFO's, aliens, alien bases on the moon, and alien civilizations. And I believe that those that come forward in this country to speak the truth face organized ridicule and possible eradication.

Several men stated that we were successful in the reverse engineering of the flying saucers and we now possess a wide variety of them. A couple of these men stated that they were concerned that our government would stage a False Flag Attack (an attack planned from within) using our own UFO's and claiming that we were being attacked by aliens. They would then use this false flag attack to accomplish their diabolical objectives - perhaps an excuse for the US to compromise our sovereignty and share all of our top secrets with foreign governments in the name of defeating our common enemy - the aliens.

Donna Hare had a secret clearance while working for NASA contractor, Philco Ford. In a video, she testifies that she was shown a photo of a picture with a distinct UFO. Her colleague explained that it was his job to airbrush such evidence of UFOs out of photographs before they were released to the public. She also heard information from other Johnson Space Center employees that some astronauts had seen extraterrestrial craft and that when some of them wanted to speak out about this, they were threatened. SiriusDisclosure.com

"UFO's are as real as the planes flying over your head. It is time that the US government started to come clean on what it is all about because there are very important military and economic issues that need to be addressed". Paul Hellyer, Former Ministry of Defense, Canada.

     Wernher von Braun was THE TOP German, (later American,) aerospace engineer credited with inventing the V-2 rocket for Nazi Germany and the Saturn V for the United States. Von Braun is on record (I remember his film interview) saying that he received help with his designs by aliens! Why would he lie?

     Thanks to WikiLeaks, this email became published recently:
 Email for John Podesta c/o Eryn re *Space Treaty
(John Podesta previously served as chief of staff to President Bill Clinton and Counselor to President Barack Obama.) From Edgar Mitchell, (the 6th astronaut to walk on the moon). "Because the War in Space race is heating up, I felt that you should be aware of several factors as you and I schedule our Skype talk. Remember, our non-violent **ETI from the contiguous universe are helping us bring zero point energy to Earth. They will not tolerate any forms of military violence on Earth or in space. The following information was shared with me by my colleague, Carol Rosin, who worker closely for several years with Wernher von Braun before his death. I have worked on the Treaty on the Prevention of the Placement of Weapons in Outer Space, attached for your convenience."   *Outer Space Treaty
**Extra Terrestrial Intelligence

     Speaking of UFO's, I was in San Diego waiting for a train one night. While at the train station, a man in a wheelchair comes up to me, looks up in the sky and asks if I noticed the "moving stars". I looked up and I saw 3 or 4 stars moving together in formation. I never would have noticed the "stars" without him pointing them out to me. It was a partly cloudy night, which he said, is the only time he sees them. He never sees them on a clear night because he believed that they used the clouds to help conceal their movement. I asked the gentleman if the moving stars belonged to "them", or if they are ours. He said that he didn't know.

     Over the years, I've had several dreams about UFO's. Here's a typical one: people around me are surprised and in a bit in a panic

about the UFO which apparently landed in the distance. When I hear about it, I raise my hand and say, "they're here for me."
In another dream I was inside a spaceship looking out the window. I remember it being a very smooth ride.

When I was filming my video documentary about the Great Pyramid, for one scene, I put my camera on the tripod, turned it on and walked into the shoot with the pyramid in the background. As I was walking alongside it and looking up at it, I got a bit dizzy and felt like I was going to lose my balance. It was an unusual feeling. I remember looking down and asking myself, "what's wrong with me? I can't even keep my balance".

I decided to retake the shot and everything felt normal the second time. No unusual feelings, no dizziness. But, later that night when i was looking at the video I shot that day, I was shocked to see something really weird hovering, maneuvering and changing shape directly behind me. I asked my assistant if that was a UFO of some sort. She definitely agreed and said that it was a UFO (or some other Unexplained Aerial Phenomena). This is what made me dizzy and unbalanced. I left the shot in my video documentary and is on my website..

Anyhow, getting back to "Unacknowledged"... Dr. Greer also discusses the massive suppression of technology and destruction of new products and patents relating to the development of free energy. He claims that we are now 100 years behind because of this suppression. The new patents and resulting products would destroy the large industrial money-making complexes like Big Oil.
"We are sitting on extraterrestrial technology that can change the world forever." Dr. Steven Greer

When you think about the incredible advancements in technology during the past 10 years (laptops, cell phones, internet, etc., etc.) and compare that to the lack of advancement in the common automobile (we're still using the gasoline-fired internal combustion engine that was developed shortly after the American Civil War around 1861). I totally agree with Dr. Greer's' comment that we are decades behind.

I can't also help to believe that there are cures for certain types of cancers that may also have been suppressed. I've got a close friend who feels very strongly about this. If you think about it, cancer is a big business. Lots of money is made from people who need expensive cancer treatments. I totally forget the number that I was told, but it could have been something like: one new cancer patient equals $750,000 in new revenue. Big pharma, along with many others, have a lot to lose if a cure for cancer is found.

I believe that as we move forward in this new age, more and more truths will be exposed and masses of people will start to open their eyes and their minds.

I totally agree with Michael Ellner's popular quote:

"Just look at us, everything is backwards. Everything is upside down. Doctors destroy health, lawyers destroy justice, universities destroy knowledge, governments destroy freedom, major media destroys information, and religions destroy spirituality."

www.ingramcontent.com/pod-product-compliance
Lightning Source LLC
Chambersburg PA
CBHW052327220526
45472CB00001B/299